£20.00
n/c

Appraisal and repair of building structures

Introductory guide

Appraisal and repair of building structures

Introductory guide

Edited by R. Holland,
B. E. Montgomery-Smith
and J. F. A. Moore

 Thomas Telford, London

Published by Thomas Telford Services Ltd, Thomas Telford
House, 1 Heron Quay, London E14 4JD

First published 1992

ISBN: 0 7277 1663 8

Typeset in Great Britain by MHL Typesetting Ltd, Coventry

Printed in Great Britain by Redwood Press Limited, Melksham, Wiltshire

Foreword

The past few years have seen a significant change in attitudes to the built environment. People — including in response to popular pressure, decision-makers — no longer take it for granted that a new building will be an improvement on the one that it is intended to replace. The number of engineers involved in work on existing buildings has therefore increased considerably. However, engineers have, in the main, been educated and trained in the art of designing new structures and many are unaware that although the scientific principles do not differ the application of those principles to an existing structure is not as simple as it might at first seem.

The former Structural Engineering Board (now the Structural and Building Board) of the Institution of Civil Engineers considered that sources of guidance regarding the appraisal and repair of building structures were dispersed too widely. It therefore brought together a number of engineers with considerable experience in the subject who, in turn, suggested that a series of guides should be produced by Thomas Telford Services which would draw existing guidance together. Specialists were commissioned to deal with subjects such as reinforced concrete and foundations, but it was felt that a guide was also required to set out clearly the approach which should be taken to the appraisal and repair of building structures in general.

A drafting committee was therefore set up both to write this general guide and to advise the publishers on the appointment of specialist authors. Although members of the committee have individually taken responsibility for the chapter(s) which bear their names, all have had to run the gauntlet of approval by their peers to ensure even treatment of the various subjects.

During the editing of the final drafts we were saddened to hear of the death of Bill Curtin. Bill's excellent work on the restoration of the Albert Docks in Liverpool and his experience as a writer of technical books were the reasons for his membership of the drafting committee. The speed with which the first drafts of his chapters were completed was astounding and the good-natured way in which he took and gave criticism helped all the other members of the drafting committee to swallow our pride and see things in a better perspective.

I would like to record my thanks to the members of the drafting committee, all of whose services have been given without remuneration.

Robert Holland
January 1992

Contents

1

Introduction

1.1. Scope

This series of appraisal and repair guides is intended to provide the practising engineer in the United Kingdom with concise yet comprehensive guidance regarding the best procedures to adopt, together with detailed information regarding the specific materials likely to be encountered. This volume — the introductory guide — deals with subject matter common to most appraisals while the supporting guides deal with specific forms of construction such as foundations, cladding and fixings and with the more commonly used structural materials.

The supporting guides are written by specialists in the subjects concerned while the chapters of this introductory guide have been written individually by members of the committee tasked by the Institution of Civil Engineers with producing the series.

Although there is a definition of structure which in relation to buildings is all embracing, the intention has been to deal only with matters on which an engineer's advice is likely to be sought. The guides therefore treat in full the familiar engineering materials such as reinforced concrete and structural steel designed and built according to established practice, but with materials such as masonry and timber the aim has been to concentrate on strength, stability and durability and other engineering matters, and not to deal with those properties and uses of the materials which would generally be regarded as the responsibility of an architect or building surveyor.

It has been assumed that the engineer using these guides is conversant with building terminology and will already have some knowledge of the principal causes of deterioration.

1.2. Using the guides

This introductory guide has been designed to be used as a checklist. Section 1.4 provides an outline procedure which takes the reader through all the stages of an appraisal as well as providing references which will bring the other chapters of the guide into context.

To ensure conciseness, information which already exists in published form has only been repeated in full where it is considered that by doing so there is some real advantage to the reader, e.g. in limiting the number of publications an engineer will need to take to site with him. Adequate references are, however, provided at the end of each chapter to what the authors consider to be the most useful and relevant publications.

1.3. Purpose of the appraisal

It is important to establish initially why a structural appraisal is to be carried out, as this will affect the amount of data to be collected and analysed before the final report is prepared. The principal reasons for carrying out an appraisal are

- serious deterioration — immediate advice, in advance of a full appraisal report, may be required if premature collapse is likely or where collapse has already occurred

- significant deterioration — the extent to which advice on maintenance as well as remedial works is required should be established

- defects in design or construction — the engineer should consider the reputations of the parties involved by being objective at all times and allowing the facts to speak for themselves.

- accidental damage and resulting collapse — it is important to establish immediately the extent to which there is danger

2

from further collapse and to render the structure safe and, as far as is practicable, weathertight

- proposed change of use

- for purchase — it should be established quite clearly whether the sole objective of the appraisal is to help the client to decide whether or not to purchase. Where a quick response is required the client should be advised of the inherent risks

- for sale — the liability of the engineer to prospective, and probably unknown, purchasers should be borne in mind

- for insurance or legal purposes

1.4. Establishing the brief

Only in an emergency situation should an appraisal be commenced without a brief having being agreed with the client. It will often be prudent to limit the first brief to that required for an initial appraisal. A detailed brief for the full scale appraisal can then be drawn up when the results of the initial appraisal are available. Many clients may need some help from the engineer in preparing the brief. The points to be covered in any brief are

- to establish the purpose of the appraisal (see section 1.3)

- to state in general terms how the appraisal will be carried out, detail the investigations to be undertaken and what they seek to establish

- to state what circumstances will require the engineer to seek further instructions from the client

- to define the line of responsibility especially when the appraisal is carried out jointly or on behalf of more than one client

- to confirm what other experts or testing houses are to be employed and how they are to be remunerated

- to establish the time and funds available and the limitations which these will place on the reliability of the appraisal

- to agree the extent of investigations which are needed to support the appraisal and to give an estimate of costs together with details of how payment is to be made.

1.5. Appraisal procedure

Although the approach to the appraisal of a simple structure and that for one of more complexity will differ considerably, the following headings will provide a comprehensive checklist or outline procedure which is summarized in Fig. 1.1.

- The brief — guidance is given in section 1.4

- Safety — it is important to establish at the start the extent to which investigators will be in any danger from total or partial collapse of the structure or from other sources

- Access — guidance is given in Chapter 9

- Initial appraisal — guidance is given in Chapter 2, Chapter 3 includes advice on signs of structural distress

- 'Desk-top' study — the time should then be taken to consider the information obtained so far, to review the brief and to plan the remainder of the appraisal in detail. Although the initial appraisal may have indicated the general direction in which investigations should proceed, it is nevertheless prudent not to discard alternatives until all reasonable objections to

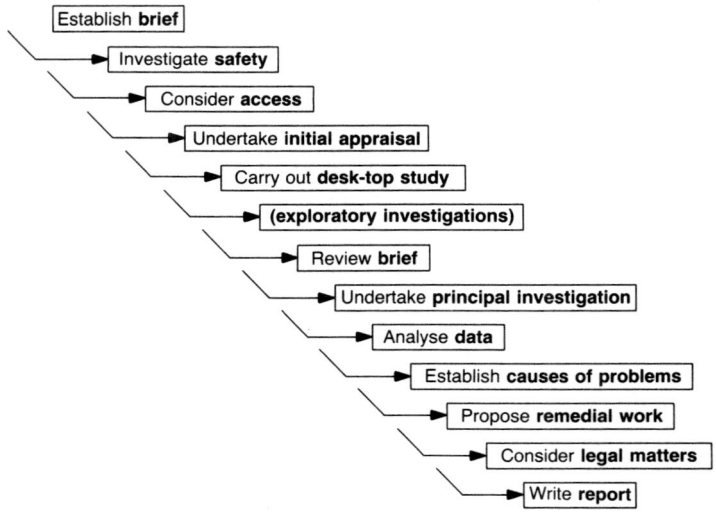

Fig. 1.1. Appraisal procedure

them have been validated. This may require further background information and it may be necessary to consult others with particular experience in the appropriate field. The supporting guides in this series give information relevant to specific materials and forms of construction and Chapters 13 and 14 of this volume deal with unusual materials and structures, and historic buildings

- Exploratory investigations — before a programme of extensive investigations or tests are embarked upon it is normally prudent to carry out a series of exploratory tests, the object of which would be to highlight access and other difficulties which may not have been foreseen initially, to confirm any previous cost and time estimates and to establish that the tests envisaged are the most appropriate

- Review of the brief — if the information gathered so far indicates that changes to the brief will be required, the client should be contacted for his instructions

- Principal investigations — the principal investigations can then be put in hand. Guidance on methods of investigation is given in Chapter 4 and on instrumentation and monitoring in Chapters 10 and 11. Chapter 12 deals with the load testing of structures

- Analysis of the data — Chapter 6 gives guidance on assessing the strength of existing structures and chapter 3 of the Institution of Structural Engineers' publication *Appraisal of existing structures* describes the process of assessment in detail

- Establishing the causes of failure/deterioration — guidance on causes of failure can be found in Chapter 5

- Remedial measures/maintenance required — Chapter 7 gives advice on remedial measures

- Legal and other non-engineering matters — Chapter 8 gives advice on legal aspects and Chapter 14 deals with the special requirement for historic buildings

- Report — Chapter 15 gives advice on writing the report

1.6. Making recommendations

Any recommendations that are made must concentrate on addressing the problems as seen by the client and be in a language which the client's commissioning agent will readily understand. It follows that a report for a resident householder should explain unfamiliar technical terms whereas one for the building department of a large organization will assume a higher level of technical knowledge in the reader.

Select bibliography

1. Building Research Establishment. *Structural appraisal of existing buildings for change of use.* BRE Digest 366, 1991.
2. Currie R. J. Towards more realistic structural evaluation. *The Structural Engineer,* **68** No.12, 1990.
3. Greenspan H. F. *et al. Guidelines for failure investigation.* American Society of Civil Engineers, New York, 1980.
4. Hollis M. *Surveying buildings.* Surveyors Publications, London, 1986.
5. Institution of Civil Engineers. *Repair and renewal of buildings.* Thomas Telford, London, 1983.
6. Institution of Structural Engineers. *Appraisal of existing structures.* IStructE, London, 1980.
7. Institution of Structural Engineers. *Guide to surveys and inspections of buildings and similar structures.* IStructE, London, 1991.
8. Mika S. L. and Desch S. C. *Structural surveying.* Macmillan Education, London, 1988.
9. Richardson C. A guide to structural surveys. *Architects' Journal* reprint, 1985.
10. Robson R. *Structural appraisal of traditional buildings.* Gower Technical Books, Aldershot, 1991.

2

Initial appraisal

2.1. Purpose and objective

The purpose of the initial appraisal is to gain a feel for the nature
and extent of the work to be carried out and to assist in agreeing
the final brief with the client. The objectives should be

- to prepare an agreed brief for the full appraisal
- to gain an understanding of the main problems likely to be
  encountered
- to obtain sufficient information to forecast the resources needed
  for the full appraisal
- to plan any specialist services

2.2. Preliminary work

Before visiting the site, pre-planning and desk-top investigation
should be undertaken to gain familiarity with the history of the
building and its surrounding area, the types of building to be found
in the neighbourhood and any information or special
circumstances such as ground subsidence or settlement in the
locality.

The local authority's building control department should be
contacted for permission to examine any records they may hold
of the particular property, the original design calculations or

structural drawings and any information they may be able to provide regarding potential problems in the area such as settlement or unusually severe exposure.

The British Geological Survey may be able to provide some details of the geological strata of the area and, if necessary, the relevant geological maps and records should be studied. If the site is in a possible mining area the surveyor for British Coal should be asked for a report on the mineral records he holds, details of any recorded past mine workings in his possession and any proposed plans there may be for mineral extraction in the future. Enquiries should be made regarding other minerals such as limestone or salt which may have been extracted.

The client may have records, drawings or original structural calculations of the building to be investigated, details of alterations or additions to the original building and records of maintenance or repair works that have been carried out.

2.3. Equipment

The basic equipment listed below will be required on most initial appraisal visits and can be supplemented from the additional list where there are specific requirements.

- Basic equipment
  - binoculars
  - camera with flash and films, powerful torch
  - tape measure and folding rule
  - crack width measuring device
  - mirror
  - spirit level
  - plumb line
  - notebook and pencils
  - sketch plans and elevations

- Useful additional equipment
  - penknife
  - screwdriver
  - claw-hammer, club-hammer
  - floorboard saw
  - calipers
  - bottle of phenolphthalene and spray bottle
  - electric drill, extension cable and adapters borescope

- o telephoto and wide-angle camera lenses
- o damp meter
- o cover meter
- o shovel
- o pick
- o folding stepladder
- o soil penetrometer
- o plastic bags, ties and labels
- o protective clothing — overalls, goggles, hard hat, wellington boots, gloves

2.4. First impressions

Once the pre-planning is underway the initial site visit can be made. Time should be taken to walk around the building under examination together with the buildings in the surrounding neighbourhood. Each elevation of the buildings should be examined in turn to gain an overall impression as to whether the vertical and horizontal features are straight and true. Obvious visual signs of movement or repair work to buildings in the area should be noted and photographed in order to form an overall picture of the stability of the ground likely to exist on the site.

People met during the visit may have useful information about the building. Determine how long they have known the premises and if they recall any problems with the building or work having to be done. A personal anecdote may provide an invaluable clue to the investigations.

2.5. External examination

The building to be appraised should then be examined in detail from ground level using a pair of good binoculars. A sketch and photograph of each elevation in turn should be made and all signs of movement should be recorded as they are observed. Are the ridge lines straight and true? Are all the windows and door openings perfectly square? Do any of the façade walls lean out of plumb or bulge at a particular point? Are there any cracks visible or signs of old cracks that have been repaired?

The direction of any slope to the ridge lines, window openings and the like that is apparent should be noted, as should the extent and pattern of any cracks, the direction in which they run and which way they open up. Any obvious external factors that could have an influence on the building, now or in the future, should also be noted, e.g. trees reaching maturity or stumps indicating a recent felling, new extensions, paved areas or drainage runs, redevelopment of an adjacent site or realignment of a nearby roadway.

2.6. Internal examination

After a thorough and complete examination of the exterior of the building a similar inspection should be undertaken inside. A sketch of each floor and, if appropriate, each room is required. Note whether or not the floors and ceilings are straight and level, the doors and openings truely square, and record any cracks, either open or repaired, in the walls or ceilings. Photographs should be taken including, if scale is important, a reference object such as a person or a ruler.

As each floor is examined in turn corroboration of the information obtained — both initially regarding the structural form of the building and subsequently during the external examination — should be sought. Roof voids, basements and cupboards must be inspected with particular care as in such locations signs of repairs and defects are less likely to have been hidden or re-decorated. Where there is no obvious means of access a note on the sketch should state that the particular area has not been examined.

It is essential that the way the building is structured is determined correctly. Assumptions should not be made from superficial observations without checking thoroughly. For instance the direction of the timber boards is not always perpendicular to the supporting joists. Floors may have been falsely built up to hide a sag in the inadequate original construction.

It is essential to determine whether the building is a framed structure or one utilizing load-bearing walls. If it is framed the framing principle and the individual members should be shown on the sketch plans. If it is load-bearing the structural walls should be indicated. If these do not line up with the walls on other floors, this should be made clear on the sketches.

The marked-up plans should be examined in detail to ensure that each load-path is complete. Any load-path or structural element that cannot be verified without opening up the fabric of the building should be noted. These further disruptive investigations will need the client's approval and are dealt with later in this guide.

Any obvious internal factors that could have an influence on the building, now or in the future, should be recorded, e.g. new heating or air-conditioning systems, double-glazing or insulation, heavy machinery or concentrated loadings, new walls or computer floors.

2.7. On-site assessment

Before leaving the site the sketch plans of the building should be studied carefully to compare any signs of distress observed with structural form. It is useful at this stage to try to determine what mechanism has caused the distress. Chapter 5 gives useful guidance. In some cases the causes may be obvious from the outset, such as settlement at one end of the building: the window openings, floors and roof ridge will all slope in the same direction; cracks in the longitudinal walls will open at the top. More often the mechanism is not so easily deduced and it may be one of several possibilities.

Great care is needed to distinguish between movements caused by shrinkage or expansion of the building materials and those caused by foundation settlement or subsidence. Junctions between differing materials should be carefully assessed. Some materials expand with time while others shrink, for instance where clay bricks and concrete blockwork occur together in cavity wall construction. Modern heating standards often significantly reduce the moisture content of timber resulting in differing orders of shrinkage with and across the grain. Each possible mechanism which could have caused the distress should be considered before deciding what further investigation, if any, is required to determine the correct cause and to choose the appropriate remedial action.

Attention should be drawn to access difficulties that will be encountered later on, what plant and equipment will be required, what specialist services will have to be arranged and particularly to any problems regarding safety that may arise.

The initial appraisal report should be as full a picture as possible of the building's structure, its problems and their probable causes, together with sufficient data to plan and cost any further investigation which may be necessary.

Select bibliography

1. Brunskill R. W. *Illustrated handbook of vernacular architecture*. London, Faber & Faber, 1987.
2. Building Research Establishment. *Common defects in low-rise traditional housing*. BRE Digest 268, 1988.
3. Building Research Establishment. *Site investigation for low-rise housing — the walk over survey*. BRE Digest 348, BRE Garston, 1989.
4. Building Research Establishment. *Climate and site development*. BRE Digest 350, BRE Garston, 1990.
5. Institution of Structural Engineers. *Appraisal of existing structures*. IStructE, London, 1980.
6. Institution of Structural Engineers. *Guide to surveys and inspections of buildings and similar structures*. IStructE, London, 1991.
7. Smith L. *Investigating old buildings*. Batsford, London, 1985.

3

Signs of distress

3.1. Introduction

The obvious signs of distress, repair, alteration or maintenance
to a building will already have been noted and recorded during
the initial appraisal described in the previous chapter. Recognizing
the less perceptible signs of trouble in a building is akin to detective
work. Clues have to be collected assiduously and each investigated
until their causes and effects have been established with certainty.
One clue that does not fit the general pattern of conclusions must
not be overlooked or discarded. On further investigation it may
prove to be the essential factor in determining the real cause of
a problem. Further investigation of a rogue clue often entails
serious disruptive opening up of the building's fabric. If this is
outside the brief that fact must be highlighted in the initial report
and the further work needed must be explained in detail. The
client must make the decision and be aware of the risk in not
following the investigation through to an ultimate conclusion.

No exhaustive list of clues can be drawn up to cover every
situation, but a useful basic checklist of signs of distress is given
in the Institution of Structural Engineers' *Appraisal of existing
structures* (9 in bibliography at the end of this chapter).

3.2. Use of one's senses

To uncover the less obvious clues it is essential to use all one's
senses, except of course taste — that could be fatal!

- Touch — springiness in a floor when jumping up and down, knocking on partitions to locate solid areas, etc.

- Smell — the smell of damp or the distinctive smells of dry rot, new paint or blocked drains, etc.

- Sound — the hollow sound of boss rendering when tapped, the change in note when a hammer is dragged over brickwork or a chain dragged over screeded floors, even creaking when the wind blows

- Sight — Visual clues normally relate to movement of one sort or another, or distress in or deterioration of a particular material

3.3. Signs of movement

The more obvious signs of movement include originally vertical or horizontal features no longer being so, cracks both new and old and the pattern they exhibit, skew openings at doorways and windows and excessive deflexion of floors and beams. However, previous attempts may have been made to deal with such problems in a cosmetic way by, e.g. by adjusting architraves. It should also be borne in mind that some defects may have existed for many years without detrimental effect — the campanile at Pisa, for example, was already leaning out of the vertical during construction. While some cracks may indicate that movement had taken place many years previously in order to relieve excessive built-up stresses, others may still exhibit movement in response to changes in temperature.

Less obvious indications of movement also need to be noticed and explained. A step in a brick wall at the damp-proof course may suggest expansion without subsequent contraction. Wrinkled wallpaper in the corner of a room may be hiding a settlement crack. A gap behind a precast concrete beam on its seating may result from a cycle of expansion and contraction, slowly edging the beam towards failure.

3.4. Signs of deterioration

Visual clues will also be found that suggest deterioration of the materials or elements forming the fabric of the building.

Cracking in pre-cast concrete cladding — the engineer must establish why it happened

- Concrete may show signs of cracking, crazing, spalling and rust staining, excessive deformation, buckling or bowing. The reason for such deterioration must be established

- Steel may show signs of delamination, corrosion or local deformation, even to the extent of perforation. Bolts may have been removed from connections or a truss member cut to accommodate services. Missing bolt or rivet heads may indicate distress

- Timber may show signs of rot or infestation, warping or buckling, splitting or crushing

3.5. Signs of alteration or repair

Illogicalities in a building's structure, such as columns that seem uncomfortably thin or surprisingly thick, walls that appear to be load-bearing on one floor but are missing on another, or window fenestrations in unusual juxtaposition with those of adjacent floors all need investigation and explanation. Has the structure been altered, a column or wall removed or additional floors added? Is the building still being used for its original purpose? Is it now being subjected to significantly higher loadings?

In old masonry buildings a common rule of thumb when alterations were being undertaken was to assume that the walls and foundations would probably carry an extra 20% and the floors and beams an extra 10%. The current loadings in the building should be checked to see if these factors have been significantly exceeded.

Alertness to detail should trigger an enquiring mind to ask questions. Inside a building one should question any area of new finishes, differing ceiling coverings, plaster decorations or skirting boards, or maybe an incongruous false ceiling or wall newly decorated, to establish whether these have been applied to conceal a defect or superficial repair. The provision of central heating in an old building may have significantly reduced the strength of timber joists through notching.

Outside a building one should question a slight change of colour of pointing or a patch of different brickwork, one façade of a building renewed with a skin of brickwork or rendered over. The balance of a façade may have been subtlely changed, a parapet removed or a window enlarged. These may conceal a defect, repair, unsatisfactory alteration or maybe a change of use of the building.

If the building has been extended sideways or upwards additions are often recognizable because they have been carried out using different materials. An extension may not have been satisfactorily married to the existing structure with adequate provision for differential movements and prevention of water penetration.

Major openings which have been formed in the original façade to give access to an extension may exhibit signs of movement or cracking.

Additional floors impose vertical and (more difficult to

determine) lateral loadings on the existing structure which it may not be capable of sustaining.

Repairs or alterations cannot necessarily be expected to have been executed honestly and competently, i.e. their effect may not be satisfactory. A deliberate attempt to conceal an inadequate repair, alteration or inherent defect is often the most difficult to detect.

Select bibliography

1. Building Research Establishment. *Floor screeds.* BRE Digest 104, BRE, Garston, 1972.
2. Building Research Establishment. *Calcium silicate (sandlime, flintlime) brickwork.* BRE Digest 157, BRE, Garston, 1973.
3. Building Research Establishment. *External rendered finishes.* BRE Digest 196, BRE, Garston, 1976.
4. Building Research Establishment. *Estimation of thermal and moisture movements of structures.* BRE Digests 227−229, BRE, Garston, 1979.
5. Building Research Establishment. *Fixings for non load-bearing cladding panels.* BRE Digest 235, BRE, Garston, 1980.
6. Building Research Establishment. *Low-rise buildings on shrinkable clays.* BRE Digests 240−242, BRE, Garston, 1980.
7. Building Research Establishment. *Surface condensation and mould growth in traditionally built dwellings.* BRE Digest 297, BRE, Garston, 1985.
8. Building Research Establishment. *Damage to structures from ground-borne vibration.* BRE Digest 353, BRE, Garston, 1990.
9. Institution of Structural Engineers. *Appraisal of existing structures.* IStructE, London, 1980.
10. Rainger P. *Mitchell's movement control in the fabric of buildings.* Batsford, London, 1983.
11. Watts G. R. *Case studies on the effects of traffic-induced vibrations on heritage buildings.* Transport and Road Research Laboratory, Research Report 156, Crowthorne, 1988.

4

Principal investigation

4.1. Introduction

The aim of the initial appraisal (see Chapter 2) was to provide sufficient information for the direction in which the principal investigation should proceed to be determined and to identify the resources which would be required. Basic detective work was dealt with in Chapter 2 and this chapter, therefore, concentrates on the methodology of the principal investigation and highlights specific problem areas.

4.2. Preliminary work

During the initial appraisal one or more hypotheses regarding the cause of any defects which have been observed are likely to have been postulated, each of which now requires examination in detail. It may well be worth carrying out a literature search or seeking the views of a specialist firm or laboratory at this stage in order to locate instances where similar defects have occurred. A few hours spent browsing in a good technical library, such as that of the Institution of Civil Engineers or of the Institution of Structural Engineers, may well provide the lead which the investigator is seeking. Conferences covering the general subject of appraisal and repair are held from time to time and their proceedings contain many case studies. It is foolish to think that the problem under consideration is unique; conversely it is dangerous to accept uncritically another investigator's conclusions without thoroughly checking both the credentials of the investigator and the significance of the similarities with one's own problem.

4.3. The nature of the evidence

The use to which data is to be put — and the previous experience of the investigator — should determine, to some extent, the means by which it ought to be acquired. For example, it may be prudent to employ recognized specialist firms to carry out some parts of an investigation if the results are required to stand up to rigorous legal examination.

4.4. The extent of investigations

Investigation is expensive — particularly where access is difficult. While the aim should be to collect only what data is necessary, the amount required will depend upon why the data is needed. The principal reasons for requiring data are

- to dispel uncertainty or ignorance regarding the physical condition of hidden construction

- to determine the physical extent of defective areas

- to predict the residual strength of a structural element

- to test hypotheses and eliminate the invalid ones

- to provide a basis for establishing correct remedial action

- to provide a basis for estimating the cost of remedial work.

One critical method of deciding the extent of investigation required — based upon the Pareto principle of diminishing returns — is to establish what the cost of not carrying out part of the investigation might be. For example if, in a 5% random sample of the beams in a large office block, one in eight were found to be of unacceptable strength it would, in order to provide more accurate cost predictions, be worth testing a larger sample before drawing up a contract for strengthening works. Conversely, if six out of eight beams had been found to be unacceptable there would be a strong argument for any further testing to be left to form part of the remedial contract.

4.5. *Foundation failure*

Although there is a further title in this series dealing specifically with foundations, the following points are included here in order to assist the investigator.

For buildings of domestic scale it is quite common for little or no theoretical design work to have been carried out — particularly in respect of foundations. Because of the scale of the work those responsible for such work were rarely members of an engineering discipline. Practice at the time of construction may not, in any case, have been advanced enough for a proper assessment of the soil to have been made.

It should not be assumed that foundations are either to the size or depth shown on any drawings or calculations — particularly where there is doubt as to the cause of observed defects. A few confirmatory trial holes will be worthwhile.

Once the type of soil and spread of footings has been determined, a calculation of the loading should be made to determine whether this is within the normally recommended band of allowable loadings for that soil and note should be taken of any significant variation in loading (likely to cause differential settlement) or possible sudden changes in soils or underlying strata (for example due to solution channels in chalk). It is always wise to check whether neighbouring properties exhibit similar defects or whether similar defects have been rectified.

A check should be made as to whether soil strengthening or other ground engineering techniques have been used and any records of the site control of such work should be carefully examined. Evidence indicating movement of nearby retaining walls should be sought as should that of the removal of trees. Existing trees will be fairly obvious and the effect of various species on different types of ground and foundations is discussed fully in the foundations guide in this series.

Signs of trenching operations close to the foundations under consideration and the quality of backfilling should be investigated as should significant building or excavation works on neighbouring properties, local mining operations or tunnelling works. The addition or removal of a neighbouring building or other structure — or even a large amount of soil — may have caused a problem due to overlap of pressure bulbs within the mutually supportive strata.

Changes to the water table and the possibility of fractured or badly constructed drains or water supply pipes, and the leaching away of fine material in the soil which can result from such fractures, need to be investigated. Drain testing may be called for and the appropriate water company can usually be persuaded to verify the integrity of supply pipes. Monitoring of the groundwater levels may provide useful data on seasonal variations and this may be accomplished by means either of a trial pit or of a simple tube sunk into the ground — the choice depending upon the ground condition. Care should be taken to prevent anyone falling into a trial pit.

Vibration damage may be due to changed traffic conditions (even temporary diversions) or by nearby construction work (as can fractured services) and it should be borne in mind that the cause of the damage may no longer, in its original form (for example, a pile-driver) be present.

Useful documentary evidence which should be sought includes borehole information (both specific to the site being considered and to neighbouring sites), clerks of works' diaries, weather reports, the results of any tests carried out at the time of construction and photographs covering as long a period of time as possible.

In-situ testing for structures larger than domestic scale may require excavations within the building itself involving the breaking out and reinstatement of flooring. Care should be taken in such circumstances to make good any disturbed damp-proof courses.

Piled structures are much more difficult to investigate. In the first instance the actual number and size of piles should be checked against drawings and calculations and the loading on each pile or group verified. Special attention needs to be paid to the amount of settlement predicted in the original calculations and that observed and consideration given as to whether the building structure is capable of transmitting the load in the manner assumed under the settlement conditions observed. The loading of piles which can result from somewhere other than the structure under consideration should be considered. For example, the adherence of soil to the shaft of end-bearing piles — where the soil is subject to settlement or is in the process of consolidation — can increase the load carried by individual piles. Pile caps should be checked for compliance with drawings and limited information may be obtained from trial holes.

Changes in the building itself which may affect the foundations include the addition or removal of very large loads, additional storeys or the removal of supporting walls, most of which should have been noted as part of the initial appraisal during which foundation rotation may also have been noted. The extent of these effects then needs to be ascertained and their significance, if any, verified.

4.6. Overstress of superstructure

Where the initial appraisal has indicated that elements of the superstructure are overstressed, the magnitude of the overstress needs to be ascertained by calculation. However, always bear in mind that heavier materials may have been used in floors, ceilings and partitions than was intended. When the calculations support the initial assumption the signs of distress need to be carefully recorded to assist in the preparation of proposals for remedial work.

Where the calculations do not support the initial assumption it is possible that the materials are not to specification (or have deteriorated), that the path of loads through the structure is not following that intended by the designer, or that the structure may have been overloaded in the past. Material strengths may be arrived at by testing samples (homogeneous materials) by counting the number of allowable defects (timber) and by using non-destructive techniques to measure hidden factors such as reinforcement. Partition walls built tight against beams or girders may carry part of their load to other beams or girders supporting the partition and structural members which lack the appropriate stiffness may shed load to others. Previous overload is the most difficult to establish with any certainty and should only be assumed when all other possibilities have been exhausted.

4.7. Thermal or moisture movement or creep

Most of the data required to assess thermal, moisture and some other causes of movement in building structures has been conveniently brought together in *Mitchell's movement control in the fabric of buildings* and a number of Building Research Establishment digests provide further guidance relating to specific materials and forms of construction.

As different materials behave differently in response to moisture, temperature and other strain-inducing phenomena, it is important to note where there is juxtaposition of dissimilar materials and the provision, lack of provision or lack of effectiveness of joints needs to be considered.

Tall buildings experience a wide range of temperatures both diurnally and seasonally. For example, a sunny day during otherwise cold weather may cause the south facing façade of a building to expand considerably in relation to the north side. Twisting of a building (Campanile, Pisa) caused by different façades reacting to strong sunlight at different times of the day has been noted. If the joints of cladding cannot cope with such the resulting movements' forces may well be transmitted to the frame of the building. The cyclic nature of such loading is particularly significant as it can lead to fatigue failure of fixings. During the winter there will be a substantial variation between internal and external temperatures and, although most building designers are aware and make suitable allowances for the phenomena in their design, bowing of panels has been known to result from this. Assumptions regarding cyclic behaviour should not, however, be made until substantiated by monitoring.

4.8. Impact

Evidence of impact is not normally difficult to establish. Localized deformation is typical and obviously the affected area would be readily accessible. The full extent of the damage may not, however, be at first apparent as it may be spread considerably further than originally assessed. Moreover the removal of a structural member as a result of the impact may result in significant overloading elsewhere.

4.9. Chemical and biological action

Chemical action is proceeding continuously in many materials used in construction and in most cases is beneficial (for example concrete increases in strength for many years after casting due to chemical changes in the cement). Some chemical actions are entirely dependent upon the constituents of a material itself but many are in response to external influences such as moisture, temperature and other chemicals. Many of the most serious chemical actions involve moisture either as a necessary agent of change or as a means of transport as, for example, in the movement of salts within masonry. One source of damaging moisture is condensation and this has been the cause of deterioration of reinforced concrete beams in an otherwise 'dry' internal environment.

Defects in concrete due to chemical action are becoming increasingly common and include carbonation which leads to exposure of the reinforcement to corrosive elements and alkali—aggregate reaction, which leads to the breaking up of the concrete matrix itself. Particular formulations of cement have their own problems (for example, high alumina cement looses strength at certain temperatures) and the effects of some additives — hopefully removed from use when their effects become known — are not entirely beneficial (for example calcium chloride, intended as a setting accelerating agent, which is corrosive). Many chemicals affect concrete or its reinforcement (including road salt) and the presence of any material in close proximity to damaged or suspect concrete should be noted — even common household products such as milk or sugar. Further information is available in the appropriate subject guide in this series.

Masonry (the term here includes brickwork, blockwork and stone masonry) performs well if kept dry. This is rarely possible and while some grades of these materials perform extremely well in the wettest of conditions others are not at all resistant to the effects of moisture. The problems usually arise where, either through ignorance or from the removal of protection, cheaper materials are subject to continuous or repeated wetting. Most of the damage results in the shedding of the surface of the unit due either to frost action or to the formation of a layer of salts left there when the moisture which brought them evaporated. Some coating systems exacerbate this. Further information is available in the appropriate subject guide in this series.

Timber, being a naturally occurring material, has many natural enemies including several species of insect and two highly destructive fungi. Damp and lack of ventilation both encourage fungal attack to occur. A number of specialist firms who incorporate surveying and treatment into one practice are active in dealing with timber problems. Many of them also advise on defective damp-proof courses in masonry and can replace or claim to simulate such by chemical means. Most offer guarantees covering their work which are acceptable to building societies as an indication that a property is a good investment. However, the appraising engineer should not be overwhelmed by the expertise of these specialists and should satisfy himself that the cause of the defects has been correctly identified and that the nature and extent of treatment is appropriate. Further information is available in the guide on appraisal and repair of timber in this series.

Steel and other metals are subject to corrosion in various ways and in differing degrees. At one end of the scale are the 'Corten' and *some* of the stainless steels which are very resistant to corrosion. Unprotected, or nominally painted, mild steel is only suitable for dry conditions; even condensation within a ceiling void (the steel being the coldest surface there) can lead to deterioration. No protection for mild steel will last for ever, although there are many treatments which will give an acceptable life in the majority of situations found within buildings. Externally the protective coatings to steelwork need to be properly and systematically maintained and the appraising engineer will need to establish the extent to which this requirement is being met. Further information, including guidance on the evaluation of cast and wrought iron is available in the appropriate guide in this series.

Metals containing copper (i.e. brasses and bronzes) have the peculiarity of exhibiting splitting after several years of satisfactory service. As a structural material the use of these metals is mainly limited to fixings and further information is available in the appropriate guide in this series.

4.10. Poor materials and workmanship

Although poor materials and workmanship are sometimes obvious it is by no means uncommon for disreputable builders — or at least the employees of less demanding contractors — to cover up faults. Instances where such has been the case include walls not centred on footings, missing wall ties, cavities partly filled with mortar droppings and foreign matter in in-situ concrete. Although there are totally unscrupulous builders and errors in setting out and the like do genuinely occur, the majority of hidden faults occur when the builder has had to deal with a building detail which presents difficulties of access or construction because of thoughtless detailing. It follows that those parts of a building which would have been difficult to construct or where any fault would have been expensive to correct should be regarded with some suspicion.

Select bibliography

1. Addleson L. *Building failure: a guide to diagnosis, remedy and prevention.* Butterworths, Oxford, 1992.
2. Institution of Structural Engineers. *Guide to surveys and inspections of buildings and similar structures.* IStructE, London, 1991.
3. Rainger P. *Mitchell's movement control in the fabric of buildings.* Batsford, London, 1983.

5

Causes of defects

5.1. Introduction

The annual cost of repairing defects is now running at over £1000 million/year — the *need* for repair is billions. The blame for defects is often placed on the materials, but too often it is the use, or misuse, of the materials which is the cause. If the materials are basically durable and the environment and loadings are known and catered for then the defects are likely to be man-made.

A factor overlooked in the post-war building boom was that although the UK climate is temperate for people it can be hostile for building fabric. In northern Canada it freezes at the onset of winter and can remain frozen until the spring thaw. There is thus one freeze−thaw cycle a year — Britain can have several freeze− thaw cycles a day. India has the heavy rains of the monsoon followed by the dry season annually — Britain can have wet− dry cycles daily. Many building materials expand and contract with change of temperature and moisture content; if this movement is not accounted for there can be failure. To take a successful building technique, method, application, etc. from one climate and transfer it, unthinking, to another can cause problems. Building materials unprotected from weathering or hostile environment will deteriorate.

5.2. Analysis of failure

A number of analyses of building failures have been carried out to determine whether they were due to poor design, construction, materials or misuse by the occupier.

The Building Research Establishment Advisory Service study (Fig. 5.1) found that 58% of all failures were due to faulty design — design in this context is often building construction design and detailing and not necessarily engineering design; 35% of faults were due to the builder's faulty execution of the work; 12% to failure of components or materials to meet acceptable performance; 11% of failures were due to misuse by the user of the building. (There is inevitably overlap since some faults are due to multiple causes).

The BRE in its Digest 176 'Failure patterns and implications' analysed the findings of over 500 investigations of building defects and in Table 5.1 the results of the common defects in various

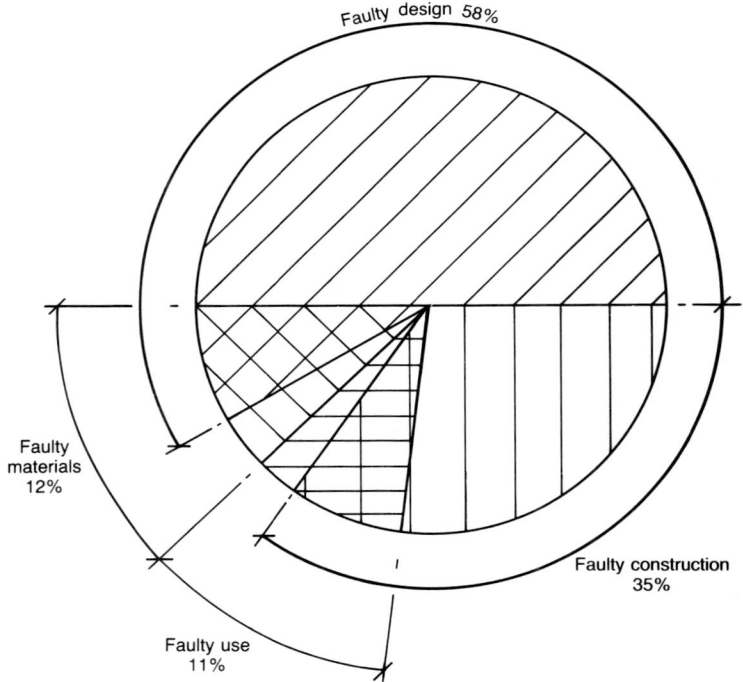

Fig. 5.1. Results of BRE's analysis of causes of building failure

Table 5.1. Failure patterns and their implications. Sample analysis by type of structure, showing the most common occurring defects

Building type	Proportion of total sample No.	%	Most common defects	Incidence in building type: %
Housing				
Council houses	64	13	Condensation	59
			Rain penetration	13
Council flats	55	11	Rain penetration	38
			Condensation	33
Private houses	65	13	Rain penetration	33
			Cracking	20
			Condensation	18
Private flats	11	2	Rain penetration	64
Non-housing				
Offices/public buildings	90	18	Floors	28
			Roofs	23
			Rain penetration	28
			Cracking	19
Factories	65	13	Cracking (mainly floors)	29
			Rain penetration (mainly roofs)	29
Schools	28	5	Roofs	39
			Floors (mainly finishes)	32
Hospitals	23	5	Floors (mainly finishes)	35
			Roofs	26
Swimming pools	13	3	Detachment of tiles	46
			Roof-space condensation	46
Universities/ colleges	21	4	} No clear pattern	
Shops	18	4		
Churches	10	2		
Other	47	9		

building types are shown. Half the defects were related to dampness (rain penetration, condensation), two-thirds of the defects impaired the functional performance of the building, while almost a quarter had minor or superficial defects (hair cracking for example) which affected the aesthetic appearance, and 11% had defects which could endanger life.

5.3. Post-war failures

After 1945 there was a desperate need to replace, renew and increase the building stock at a time when skilled labour, materials and money were scarce. This resulted in a transfer of work from the site to mass-production factories in an attempt to cut costs and save time. This represented a major departure from traditional building methods, materials and protective elements (copings, sills, etc.) and the introduction of new and untried ideas with no experience of their likely performance over their design life span. It was a challenging period, full of enthusiasm that buildings could be mass produced and assembled like cars or refrigerators. The era ended with the collapse of Ronan Point, leaving a legacy of crumbling precast reinforced concrete housing, suspect large panel systems, rotting timber frames and other disasters of 'industrialized building'. Apart from these, defects occur most frequently in low-rise, wide-span structures.

In the 50 years from the 1930s to the 1980s buildings halved in weight! Heavy pitched slate clad roofs, sturdy internal partitions, thick walls and high factors of safety have now gone.

5.4. Construction between the wars (1918–1939)

After 1918 there was no widespread abandonment of traditional practice (flat roofs which prove such a bane today were rarely built in this period). There was no serious shortage of materials, skilled labour or skilled supervision, the country had not been savagely bombed and although the troops had been promised, on their return 'homes fit for heroes' there was no strong political will to provide them, nor vigorous, organized public demand. There was therefore no building boom comparable to that at the end of the Second World War. The buildings of this era were sturdier, more robust and required less maintenance. On the other hand, fewer homes had bathrooms, internal lavatories, central heating, power points and similar amenities than in the years following the Second World War. Although there was some poor construction, particularly in private sector housing, most of the defects and problems are confined to the maintenance needed to repair fair wear and tear and to remedying faulty upgrading.

5.5. Traditional building

Buildings of the late eighteenth to early twentieth centuries did not generally have a structural frame with cladding added on. Thick, even massive, brick walls were the main load-bearing and stabilizing elements. The factors of safety of the walls and of later cast and wrought iron members were high — but often, due to ignorance, the safety factor of foundations were low. There was close integration of the two vital functions of design and construction and centuries of experience in the use of stone, timber and brickwork was enshrined in traditional details which recognized, for example the need to shed rainwater from the external envelope and drain it from roofs. There was no built-in obsolescence. The defects in these buildings mainly stem from old age, lack of maintenance, neglect and vandalism, or from alteration for change of use. Their rehabilitation now constitutes a major portion of building activity.

When dealing with building structures of this period it is necessary to take into account the methods of construction then available and the effect these had on the sequence of construction, and hence, in turn, upon the higher temporary strength required of some members during construction.

5.6. Environmental causes

There is a tendency to believe that weathering affects only the external cladding — this is not true. The interior of buildings suffer too, albeit more slowly. The principal agents of deterioration are as follows.

Temperature changes
These are mainly due to solar radiation and, to a limited extent, intermittent central heating. Steel, concrete, bricks and blocks

Inadequate cover to reinforcement in a damp cavity in an 'airy' house led to the splitting of these pre-cast concrete columns

all expand and contract with rises and falls in temperature. If this movement is not catered for in the design cracking can result and though such cracking rarely results in collapse there can be serviceability failure. Frequent cycles of heat and cold in the UK can lead to light structural elements such as parapets moving appreciably from their original position. Defects may appear in older buildings when new methods of heating and ventilation are installed.

Moisture
Moisture variations occur due to rain, snow and hail and internal high humidity from steam, condensation and dampness. Timber,

This brickwork shows the long-term damage that frost can cause

concrete, bricks and blocks can expand and contract with changes in moisture and, again, this movement must be catered for by the provision of movement joints. If the movement joints are inadequate, become defective, are bridged or in any way become ineffective, then cracking can result. Damp can rot timber, rust steel and act as a destructive agent to concrete.

Atmosphere
Corrosive atmospheres containing such chemicals as sulphur dioxide dissolved in rain-water, can corrode metals and attack roof coverings. Carbon dioxide, again dissolved in rain-water, attacks stonework, and carbonation of concrete can lead to corrosion of reinforcement. One of the commonest causes of serviceability failure of flat roofs is the gradual concentration of acids in the ponding which attacks roofing felt, asphalt and bitumen. Once the impermeable seal is broken then water leaks into the roof structures.

Air contaminated with dirt, salts or oil particles leaves deposits on buildings' surfaces. This not only affects a building's appearance but can attack metal and stone surfaces. Sea-spray around the coasts is well known for its corrosive effect on steel.

Biological agencies

Moulds and other fungi, such as wet and dry rot, and insects such as woodworm and deathwatch beetle attack timber. Moulds can form moss-like growths on flat roofs and hold moisture almost permanently. Weed growth can block gutters and downpipes and destroy or reduce the effectiveness of rain protection. Layers of dust built up on flat surfaces can also hold moisture.

Ground salts

Salts in solution in groundwater can by capillary action penetrate brick, concrete and stone. If magnesium sulphate is present there can be sulphate attack on these materials.

Natural disasters

Floods, exceptional winds, earthquakes, etc. are all obvious causes of deterioration.

Change of use of neighbouring buildings

Water table changes, removal or introduction of trees, raising of ground level, vibration of subsoil and similar ground condition changes can result in differential settlement which may lead to cracking of the fabric and structure.

5.7. Defects in materials

Any product made with fired clay such as bricks, tiles and some mosaics expand continuously but at a decelerating rate on removal from the kiln. Products made with cement such as precast, in situ concrete and concrete blocks will shrink irreversibly. To attach expanding brick slips to a shrinking concrete beam with an inflexible adhesive, for example, is to invite trouble.

Materials have differing coefficients of thermal and moisture expansion and when they are used in combination it is necessary to accommodate the varying movement. The air temperature in Britain can change by more than 20°C between day and night and more than 50°C between summer and winter. The external surface of a building due to heat build-up can show even greater temperature changes. The surface temperature of brickwork can change one-and-a-half times more rapidly than the air temperature. Black exposed surfaces, particularly when backed up by insulation, can experience double the increase in temperature of the air. Aluminium or copper sheet cladding fixed to concrete or timber will expand more than the supporting structure and if it is restrained from differential movement can buckle, crack and open up at joints. Rain-water ingress inevitably results.

Non-ferrous metals such as aluminium, copper, phosphor-bronze and others in contact with steel, particularly externally, can initiate electrochemical corrosion.

There have been serious serviceability failures with some modern materials. Joint sealants, for example, have broken down under exposure to ultraviolet light from the sun within five years and polystyrene in contact with PVC can 'drain' the plasticizer from it making the PVC brittle and causing the polystyrene to shrink away.

One should, however, retain a sense of proportion with regard to defects due to materials — reference to Fig. 5.1. shows that they only account for a small proportion of defects.

5.8. Construction faults

Although a construction defect may occasionally be found in an old building it is in the post-war stock that the majority will be found. The causes are complex. Most are due to falling standards of workmanship, inadequate understanding of how non-traditional materials should be worked and installed, together with poor supervision; although genuine mistakes are sometimes deliberately hidden so as to avoid the cost and embarrassment of remedial work. In detecting this kind of defect, therefore, it is almost as important to understand human nature as it is to understand engineering principles.

5.9. Design faults

'Design' in this context includes detailing (the preparation of working drawings for issue to the construction site). Although there are design faults due to calculation mistakes, misapplication of codes of practice, lack of full appreciation of structural analysis and underestimates of loadings, most defects are caused by poor detailing. Typical engineering design faults include

- lack of full appreciation of structural stability and robustness, omission of tying, lateral restraints, etc.

- inadequate attention to structural joints and connections

- no overall responsibility for structural assessment of hybrid structures where separate elements are designed by separate specialists

- incorrect choice of materials for exposure conditions.

- inaccuracies in assessment of loadings, provision of reinforcement, welding, etc.

- designing foundations to limit bearing pressure and not checking possible differential settlement and, for example, frost heave or the effects of shrinkable clays

- uncritical acceptance of computer output

- unthinking compliance with codes of practice in situations where a particular clause is inapplicable

- 'tunnel vision', for example, designing a retaining wall on

Poor detailing restricting thermal movement which led to the damage of this pre-cast concrete cladding

The patchwork of uneven degradation of this post-war brickwork was due to a combination of design faults: wrong bricks for the degree of exposure; inadequate capping allowing water in the top of the wall; and recessed pointing

a slope with adequate resistance to overturning, yet within an area of slip circle failure

- failure to appreciate that structural materials expand and contract and omission of movement joints

- inadequate specification of workmanship and tolerances

- production of complex and practically unbuildable details

- inadequate supervision of construction

- failure to carry movement joints through all materials in the plane of the joints

- adding insulation (or other environmental alteration) to existing buildings without appreciation of the consequences

- failure to check that assumptions made during design are still valid when construction gets underway.

5.10. User faults

Building owners are not generally experts on construction design and technology and can make serious mistakes. Common faults include the following.

- Overloading — office floors can be overloaded when used as stores for heavy, closely spaced, tall filing cabinets and safes. DIY enthusiasts might insert heavy water tanks into domestic trussed rafter roofs. Additional services hung from a roof structure can overload it.

- Alterations without structural design checks — for example truss members cut out of trussed rafters to insert water tanks; soffits of beams cut away to increase head room; joists rebated to allow installation of new services; load-bearing walls opened up and even removed to permit easier access or communication, cutting of deep horizontal or diagonal chases in load-bearing members; use of inappropriate material in repair such as unnecessarily hard and brittle mortars; removal of supports.

- Lack of maintenance — this has been a serious problem due to underspending, throughout the 1980s particularly by local authorities, on routine maintenance. External timber and steelwork have not been periodically painted, roofs have been allowed to leak, cracked and spalled concrete has not been repaired.

- Lack of periodic inspection — MOT checks on cars are compulsory, after three years and regular inspection of plant and equipment is commonplace, but lack of inspection of structures is normal since owners expect, perhaps understandably, that the structure and building fabric will somehow last for ever. BRE recommendations for frequency of inspection are given in Table 5.2.

- Unskilled and cosmetic repairs — after fire or other accidental damage, building owners will often clear up the mess and redecorate without determining whether or not any structural damage had occurred.

- Excessive wear and tear — use of steel-tyred fork-lift trucks, installation of plant inducing impact loading, floor screeds attacked by aggressive chemical spillage.

Table 5.2. Frequency of inspection of building façades

New buildings; older buildings after major alterations and repairs	Initially and then annually for first five years
Parts of all buildings where risk of damage by vandals or vehicle impact	Annually
Parts of buildings subject to severe exposure conditions, including all work above ninth floor of high structures	Every third year
Parts of all buildings over public areas	Every third year and after severe gales, heavy snowfall or other extremes of weather
Other buildings, and other parts of buildings	Every five years, or when repainting in the categories above is being scheduled if more frequent*
Parts of buildings known to contain latent defects	As appropriate to the nature and seriousness of the defect

* There are advantages in inspecting for defects during the inspection of the exterior of a building preparatory to repainting; warning signs of latent defects will not have been obscured by recent paintwork; making good of work opened up for inspection can be included in the repainting contract; any special provision needed for access can serve both purposes.

5.11. Classification of defects

Classifications can be subjective and different investigators may classify the same defect as 'minor', 'slight', 'moderate' or even 'very severe'. A good objective attempt has been made by the BRE in its Digest 251 *'Assessment of damage in low-rise structures'*, where the classification is based on ease of repair of visible damage, not on crack width alone (13 in bibliography). Categories based solely on crack width previously formulated by others have been abandoned because such measurements will often not produce a clear evaluation of the scale of damage. Also such a limited classification is not based on cause or possible worsening of the defect.

Table 5.3. Classification of damage to walls

Category of damage	Degree * of damage	Description of typical damage	Approximate crack width: mm†
0	Negligible	Hairline cracks of less than about 0.1 mm width are classified as negligible	up to 0.1
1	Very slight	Fine cracks which can be treated during normal decoration. Perhaps isolated slight fracturing in building. Cracks rarely visible in external brickwork.	up to 1
2	Slight	Cracks easily filled. Re-decoration probably required. Recurrent cracks can be masked by suitable linings. Cracks not necessarily visible externally; some external repointing may be required to ensure weathertightness. Doors and windows may stick slightly	up to 5
3	Moderate	The cracks require some opening up and can be patched by a mason. Repointing of external brickwork and possibly a small amount of brickwork to be replaced. Doors and windows sticking. Service pipes may fracture. Weathertightness often impaired.	5 to 15 (or a number of cracks up to 3)

Category of damage	Degree * of damage	Description of typical damage	Approximate crack width: mm†
4	Severe	Extensive repair work involving breaking-out and replacing sections of walls especially over doors and windows. Window and door frames distorted. Floor slipping noticeably, walls leaning or bulging noticeably, some loss of bearing in beams. Service pipes disrupted.	15 to 25 but also depends on number of cracks
5	Very severe	This requires a major repair job involving partial or complete re-building. Beams lose bearing. Walls lean badly and require shoring. Windows broken with distortion. Danger of instability.	Usually greater than 25 but depends number of cracks

* It must be emphasized that in assessing the degree of damage account must be taken of the location in the building or structure where it occurs, and also of the function of the building or structure.
† Crack width is one factor in assessing category of damage and should not be used on its own as direct measure of it. Local deviation of slope, from the horizonal to vertical of more than 1:100 will normally be clearly visible. Overall deviations in excess of 1:150 are undesirable.

The BRE recommend three broad categories of damage as a start to defining degree and suggest 'aesthetic', 'serviceability' and 'stability'. The first group (categories 0, 1 and 2, in Table 5.3) affects only the appearance of a building. The second group (categories 3 and 4) covers cracking and distortion. The third group (category 5) is where there is an unacceptable risk of partial or complete structural collapse.

Select bibliography

1. British Standards Institution. *Code of practice for dead and imposed loads.* BS 6399 Part 1, BSI, London, 1984.
2. Building Research Establishment. *Condensation.* BRE Digest 110, BRE, Garston, 1972.
3. Building Research Establishment. *Wind environment around tall buildings.* BRE Digest 141, BRE, Garston, 1972.
4. Building Research Establishment. *Condensation in roofs.* BRE Digest 180, BRE, Garston, 1975.
5. Building Research Establishment. *Failure patterns and implications.* BRE Digest 176, BRE, Garston, 1976.
6. Building Research Establishment. *Wall cladding defects and their diagnosis.* BRE Digest 217, BRE, Garston, 1978.
7. Building Research Establishment. *Investigations of the structural condition of building populations.* BRE, Garston, 1979.
8. Building Research Establishment. *Structural appraisal of buildings with long-span roofs.* BRE Digest 282, BRE, Garston, 1984.
9. Building Research Establishment. *Dust explosions.* BRE Digest 288, BRE, Garston, 1984.
10. Building Research Establishment. *Wind on canopy roofs.* BRE Digest 284, BRE, Garston, 1986.
11. Building Research Establishment. *Loads on roofs from snow drifting against obstructions and in valleys.* BRE Digest 332, BRE, Garston, 1988.
12. Building Research Establishment. *The assessment of wind loads.* BRE Digest 346, BRE, Garston, 1989.
13. Building Research Establishment. *Assessment of damage in low-rise structures.* BRE Digest 251, BRE, Garston, 1990.
14. Building Research Establishment. *Why do buildings crack?* BRE Digest 361, BRE, Garston, 1991.
15. Institution of Structural Engineers. *Stability of buildings.* IStructE, London, 1988.
16. Institution of Structural Engineers. *Guidance notes on the security of the outer leaf of large concrete panels of sandwich construction.* IStructE, London, 1989.
17. Rainger P. *Mitchell's movement control in the fabric of buildings.* Batsford, London, 1983.
18. Urbanowicz C. *Weaponry in structural survey. 1.* Effective diagnosis of material problems and defects in construction. *Structural Survey,*

4, No. 1; 2. Common building defects and their diagnosis. *Structural Survey*, **5**, No. 2. 1986.
19. Williams C. Structural vibration. *Structural Survey*, **8**, Nos 3 and 4, 1990.

6

Assessment of strength

6.1. Introduction

The process by which the strength or load-carrying capacity of an existing structure is assessed differs from that for the design of a new structure in a number of ways. Contemporary codes of practice are rarely directly applicable to existing structures and should only be used after careful consideration of the points made in this chapter.

6.2. Age of the structure

Modern codes refer to materials and techniques which were rarely available in their present form in past decades. Because of increasing quality control and the need for economy in the use of material, the allowable bending stress for structural steelwork, for example, has increased by 25% over the past 50 years. Over the same period new methods of analysis (plastic design for structural steelwork for example) have been introduced which have enabled smaller sections to be used with greater confidence. Thus, while the allowable stresses in a material may have to be reduced for older structures, it is sometimes possible to prove that they are adequate by using more up-to-date methods of analysis.

It is possible to find out what methods of analysis and working stresses were in general use at the time a structure was designed by referring to the *Proceedings of the Institution of Civil Engineers* and

47

other publications produced at about the time the original structure was designed. A number of more recent publications bring together such information and these are referred to in the specialist guides of this series.

6.3. Load testing

Load testing is one method of determining the strength of a structure directly but as it is very expensive it should only be used where other methods are not sufficiently reliable. Guidance on load testing is given in Chapter 12.

6.4. Evidence from the structure

Measurement (for example, of deflexions) and visual inspection (e.g., of cracks) can provide evidence of the extent to which a structure is nearing its collapse load and hence the extent of additional strength, if any, which is likely to be available. In many old structures, such as masonry walls, which exhibit no obvious signs of distress it is not unreasonable to assume that load could safely be increased. On the other hand, while measurement and visual inspection may well indicate reserves of bending or compressive strength, such observations are unlikely to be helpful when dealing with shear or tensile strength as shear and tensile failures occur with little warning.

6.5. Factors of safety

Because all the evidence regarding the strength of materials, tolerances and deterioration of a structure are available to the assessor, it is possible to reduce factors of safety below those which would be used by a designer whose structure has yet to be realized. Excellent guidance on this aspect is given by the Institution of Structural Engineers in *The appraisal of existing structures* (3 in bibliography)

Select bibliography

1. Beckmann P. and Happold E. *Appraisal — a critical process of inspection and calculation*. International Association for Bridge and Structural

Engineering Symposium Proceedings, Venice, vol. 45, pp. 31—38, 1983.
2. British Standards Institution. *Code of practice for dead and imposed loads*. BS 6399 Part 1, BSI, London, 1984.
3. Institution of Structural Engineers. *The appraisal of existing structures*. IStructE, London, 1980.
4. Tassios T. P. *Physical and mathematical models for re-design of dangerous structures*. International Association for Bridge and Structural Engineering Symposium Proceedings, Venice, vol. 45, pp. 29—37, 1983.
5. Zampa F., Modena C., and Odorizzi S. *A method to assess the reliability of actual buildings*. International Association for Bridge and Structural Engineering Symposium Proceedings, Venice, vol. 45, pp. 259—266, 1983.

7

Remedial works

7.1. Major options

Choosing between repair and total or partial demolition followed
by replacement is often difficult. The issues which have to be
addressed in formulating a recommendation are listed in this
section.

Economic
No matter how carefully the appraisal is carried out there will
inevitably be difficulties encountered during renovation which
were not anticipated. These will not only result in increases in
the direct costs of materials and labour but also in disruption and
prolongation costs. Although renovation will often appear to be
the cheaper option, the economic risks are greater than in
demolition and rebuild.

Residual life
Although it may be feasible to repair a building, there can be
elements which are at present quite sound but whose expected
residual useful economic life is in doubt. Where these are parts
of the structure the engineer can advise. Where they are not, the
client's attention needs to be drawn to them and he should be
advised to seek an appropriate professional opinion.

Usefulness of repaired building
The planning of older buildings is often not conducive to efficient modern use. Storey heights, access, thermal insulation and the economic use of the site as a whole need to be considered by the client and his attention should be drawn to these factors. Particularly important is the integrity of the structure as a whole and whether it can sustain modern loadings.

Intrinsic value
Buildings of historical or architectural importance are usually protected by legislation. Others may form part of a group of buildings which, if broken up by the demolition of one of them, would lead to the local ambience being destroyed. Consideration should be given to the possibility of repairing such a building without destroying its character. In such situations it is common practice to retain the façade of a building while reconstructing the interior.

7.2. Strategy for remedial work

Standards
The standard which is acceptable to the client needs to be agreed. In its BMI Special Report 167 *Condition Surveys* the Royal Institute of Chartered Surveyors lists four general levels of condition (20 in bibliography).

- As new with the expectancy that with proper maintenance the building will provide a satisfactory standard of service

- Satisfactory, safe, with only minor deterioration which can be dealt with within existing maintenance budgets

- The building is operational but major repair or replacement will be necessary within a reasonably short period with costs outside the current maintenance programme

- Inoperable, unsafe, with a risk of immediate breakdown requiring urgent expenditure outside the current maintenance programme.

Risks and procedures
The procedures which will be adopted for the remedial works should be clearly reported and explained to the client so that he

understands the interrelationship of the parties involved and the risks which are inevitable in this type of work. He should be asked to appoint one person from his own staff with sufficient authority to make any necessary decisions on behalf of the whole client organization. It is useful if the main areas of risk are listed and provisional procedures agreed for dealing with them should they occur. For example, if a particular defect proves to be more extensive than anticipated the engineer may be given authority to remedy this using funds set aside for work occurring later in the project. This could give the client time either to raise additional funds or to consider amending the later work if he is unable to obtain the necessary extra finance.

7.3. Economics of repair

In the ideal structure all elements will last for the same time. Although no structure is ideal, there is little justification for expending resources on extending the life span of some elements when others, particularly those of greater significance, have a limited life.

Replace or restore?

Elements which have few defects can usually be restored more cheaply than they can be replaced. Elements with many defects may be too costly to restore. Replacement requires careful consideration of accessibility, facilities for lifting into position, stability and compatibility of new components with old structures. There may also be overriding aesthetic reasons for restoring rather than replacing.

Cost of materials

The cost of structural repair materials is usually a minor consideration when compared with the overall repair costs of a building. Savings brought about by reducing the specification for materials should therefore be avoided if there is any risk that this could adversely affect the repair life.

7.4. Considerations before choosing repair method

Establish cause of defects

Do not attempt a repair without understanding what is the cause of the original defect. Examples abound of entirely inappropriate treatment applied because the investigator lacked the experience or will to pursue the necessary detective work. Three areas where this fault is most common are in establishing the causes of dampness, in the use of applied coatings and in dealing with cracking. Other chapters in this guide and in the others in the series give more detailed guidance on these topics.

Understand the structure
Many older buildings have been modified a number of times, often on a piecemeal basis and such modifications may have weakened the original structure. In buildings where there is no distinct structural frame, the elements may well combine to help prevent collapse. Alternative load paths, load distribution, membrane or arch action and other factors often combine to hold buildings together which, from simple calculations, would otherwise appear to be unsafe. Cracks and other signs of movement, particularly in older buildings, may have occurred some years before and may no longer have any structural significance. However, the structure as it now exists must be fully understood before any decisions are taken. In particular movement joints, whether provided as such originally or which have developed during the life of the structure, must be catered for.

Previous reports
Care should be taken when dealing with reports prepared by others. For example, surveyors employed by building societies frequently comment on the necessity to investigate defects in structures, often recommending that a specialist firm should be employed. Such specialist firms carry out surveys in order to obtain business and often specialize only in one field. As a result, the significance of apparent defects may be magnified out of proportion — symptoms rather than defects may have been treated and serious structural faults missed altogether.

Check previous repair history
It is useful to know the extent to which previous attempts to repair have been successful. Some repair methods, particularly applied coatings, may limit the choice available for further repair.

Condensation and damp
The civil or structural engineer is rarely an expert in dealing with

the problem of dampness in buildings, particularly where differences of temperature and humidity are involved. In such circumstances it is appropriate to seek the expertise of a building services engineer. Dampness caused through faulty building details on the other hand, is the field of the building surveyor. There is, however, useful material on both subjects in Appendix 3 of B. A. Richardson's Book *Remedial treatment of buildings* (19 in bibliography). The passage of water through engineering structures such as basements is within the specialism of the civil and structural engineer and the subject is dealt with in the guides to appraisal and repair of reinforced concrete and of masonry in this series.

Site difficulties

Considerations which may affect the choice of repair methods and materials include access difficulties, fire risks, noxious chemicals and other dangers. Environmental factors include the exposure of the site and its proximity to industrial or marine environments.

Legal considerations

In some cases it will be necessary to apply to the local authority for approval to use a remedial treatment. It may also be necessary to make an application for planning consent if the treatment is likely to change the external appearance of the building or there is a possible influence on health and safety. Most local authorities are willing to give advice on the detailed application of the regulations. Party wall agreements can be complex and it will be necessary to agree apportionment of cost between the parties. Special arrangements apply in London. Extensive structural repairs to listed buildings should not be undertaken without obtaining the necessary consent from the local authority.

7.5. Structural integrity during repair

Consideration must be given to propping, shoring and temporary works at an early stage. Many older buildings have hidden strength reserves but high quality finishings can often hide poor construction — particularly with stucco and stone construction. Alternative load paths should be carefully considered in case first assumptions as to the condition of hidden parts of the structure prove to be false. Cracking can be avoided during transfer of load from an old structural element to a new one if folding wedges or jacks are used to prestress the new structure before the load is transferred.

7.6. Essential characteristics of remedial works

Buildability
Simple solutions, where possible, are always best. Materials and methods involving health hazards, difficult working conditions or fire risks should be avoided. Small-scale preliminary development trials of working methods will often be useful before embarking on an ambitious scheme of remedial work. The ideal repair material will be easy to prepare and apply to a high standard. Pre-batching of component materials by the supplier may eliminate errors.

Compatibility
Solutions should be flexible enough to deal with all expected conditions as the full extent of remedial works is unlikely to be seen until the stripping out is done. The effect of the repair on the existing structure needs to be considered. Partial underpinning, for example, while solving an immediate problem can create distress in adjacent parts of the structure or even in neighbouring premises. Mismatch of materials should be properly assessed both by checking possible chemical reactions between adjoining materials and understanding which of the many physical attributes of materials are the most important when they are placed together. For example, the relative strains of two materials, the first in the original structure and the second in the repair, may be of greater importance than the strength of those materials. In other repairs, a different set of criteria may have to be considered.

Durability
Long-term performance data are often difficult to come by and accelerated testing techniques can at best be comparative as between one material and another. Agrément certificates can be useful in assessing durability but any limitations noted on them should be read with care. However, progress has to be made in the industry and untried methods and materials could offer a more

economic alternative which the engineer may well think is worth the risk involved. It is important, however, that the client fully understands and agrees to what is being attempted. Durability is also affected by the ease with which maintenance can be carried out and the frequency necessary. Maintenance-free structures are inherently the most durable.

7.7. Choosing methods and materials

Although previous experience may dictate that there is only one economic and effective solution to a problem, it should never be embarked upon without first considering alternatives both from the point of view of effectiveness and cost. The market changes and technology is constantly developing, so yesterday's solution may have been overtaken.

Free advice
A great deal of free advice is provided by commercial organizations and trade associations. Often, however, the engineer has to probe behind the information provided because that which is not provided may be of more significance. This aspect will be dealt with in more detail in the other guides in this series.

Important physical properties of materials
This list, while not exhaustive, constitutes a checklist of physical properties of materials which may have to be taken into consideration.

- compressive strength
- tensile strength
- bending strength
- shear strength
- bond strength
- torsional strength
- bearing strength

- yield strength
- wet strength
- fatigue strength
- brittleness

- ductility
- hardness
- toughness
- rigidity

- durability
- ageing properties
- frost resistance
- abrasion resistance

o stiffness
o plasticity
o creep
o density
o modulus of elasticity
o coefficient of thermal
  expansion
o dimensional stability

o impermeability
o rain permeability
o water vapour
  permeability
o permeability to various
  gases and chemicals
o absorbency
o absorption

o ease of use
o workability
o economics of use
o handling characteristics
o 'nailability'
o 'formability'
o 'weldability'
o adhesiveness
o toxicity
o storage life
o setting time
o curing time

o fungus and insect
  resistance
o resistance to aggressive
  environments
o resistance to dampness
  or chemical reaction
o reaction to light
o chemical change
  properties
o vandal resistance

o thermal resistance
o fire resistance
o inflammability
o ignitability
o flame spread

o appearance
o opacity
o surface texture
o colour
o decorative properties
o 'cleanability'
o ease of maintenance
o frequency of
  maintenance

7.8. Choosing suppliers and contractors

Warranties and guarantees

The terms of guarantees and warranties need to be carefully read as, rather than adding to the purchaser's rights, they may in fact reduce them. Moreover, warranties which are not underwritten by a third party such as an insurance company depend for their security on the continued existence of the firm giving the warranty.

Approved contractors
There are several approval schemes in operation, such as membership of the remedial treatment section of the British Wood Preserving Association, which imply a measure of competence, experience and (in some schemes) financial stability. Experienced suppliers of materials will often provide lists of contractors whom they have found to be satisfactory. There are, however, many failures of companies (particularly in the domestic sector) undertaking remedial treatments and extreme care is needed in making a selection.

7.9. Contractual methods

There are three principal contractual arrangements for carrying out remedial works.

Using a specialist contractor
Using a specialist contractor works well when the bulk of the work fits within the specialism of the contractor appointed such as is often the case with repairs to reinforced concrete or timber.

Using a general building contractor with specialist sub-contractors
This method is applicable when several specialists are required to work on the same project or where there is a great deal of general cutting and making good to enable a single specialist firm to obtain access. Only those general contractors with experience in the repair field should be used.

Using a management contractor
There are now several management contractors who specialize in refurbishment works. Although there are additional fees to pay, the use of this arrangement can be effective in terms of cost and time where a considerable amount of repair work has to be carried out while the building is in use or where repairs have to be co-ordinated with other work such as extensions.

7.10. Contract documentation

The aim should be to specify and measure the work to be carried out as fully as possible. Unfortunately, the very nature of repair work mitigates against ever achieving this, so it is necessary to build flexibility into the contract documentation.

Bills of quantity are useful when the greater proportion of the contract can be specified in detail as they provide the most realistic basis for controlling cost. Approximate bills (the quantities, rather than the Specification, being approximate) are appropriate when the full extent of the work is not known. Schedules of rates are more appropriate to smaller and to specialist contracts as their use makes greater demands on those responsible for cost control during construction.

Drawings should be prepared as fully as possible so as to show the contractor how to deal with all situations which are likely to occur, particularly where two or more materials meet. Inevitably with this type of work, there will be a need to prepare additional drawings when problem areas are opened up and the drawing office must be prepared to provide a rapid service if costs and progress are to be properly controlled.

Specifications are more difficult to prepare for repair work than for new construction because there are no libraries of clauses available as there are for the latter. Many suppliers of specialist repair materials provide specification clauses for their own products and, suitably amended to meet the requirements of the project, these can be very useful. They should not be used uncritically however.

7.11. Supervision

Before work starts

Before any work starts it is important to prepare a schedule of condition for any buildings or structures which might be affected by the work. This should be agreed both with the contractor and with other interested parties such as adjacent owners. The specification should give some indication of the type of site organization and level of the facilities to be provided and it will be necessary to ensure that site personnel proposed by the contractor have the appropriate experience. It will also be necessary to set up an appropriate site organization for the engineer and to ensure that the engineer has handed to the contractor any information he has regarding potential hazards such as buried electrical cables, gas and water pipes, etc.

Check assumptions

Many assumptions will have been made during the design period which can only be confirmed as work proceeds. The contractor should be given guidelines as to when to call in the engineer for further instructions. It is often difficult during a survey to measure accurately by how much walls are out of plumb or bowed. Once the scaffolding is in place however, all parts of the building are more easily accessible. In the early stages it is also useful to visit the site regularly to ensure that the Specification's requirements have been correctly understood.

Check stability

Although the contractor will carry the responsibility for care of the works and compliance with statutory safety requirements, it is likely that only the engineer fully understands the structure which is being restored and it is therefore incumbent upon him to ensure that nothing is done to the building which could endanger its overall stability. If someone is injured due to a partial collapse, particularly if the collapse occurs shortly after a visit, the engineer could well find himself facing a claim for negligence.

Quality control

It will be necessary to decide the frequency and extent of material sampling to be undertaken and what tests are appropriate. Particular attention should be paid to mixing equipment, and especially to any dosing devices which may form part of such equipment and the strict observance of manufacturers' instructions where appropriate. Although the contractor will have equipment on site it is often wise, particularly with measuring devices, for the engineer's staff to use separate equipment to avoid error due to the use of damaged or faulty instruments.

Site instructions

The very nature of remedial work requires a quick response for requests for information at site. It is essential in this respect that the engineer carries a duplicate book with him by means of which he can write an instruction to the contractor on the spot to ensure that there is no misunderstanding that could lead to claims or disputes.

General observations
The visiting engineer needs to monitor the adequacy of the contractor's supervision and the experience of the work force employed. The adequacy of the protection and security of the work must be assessed as must the equipment being used. Check that refuse is disposed of properly (not under floorboards) and that no nuisance is caused. Proper and readily available hygiene facilities are important. Urine can be extremely damaging to a number of building materials.

Equipment
The following equipment can easily be fitted into a shoulder bag and will be found useful when visiting the site.

- tape measure
- plumb bob
- spirit level
- sharp penknife
- micrometer
- notebook
- hand-held tape recorder
- torch
- moisture meter
- camera and flash

7.12. Surface treatments

Surface treatments are applied to a variety of substrates to improve their performance in relation to durability; these are dealt with in the appropriate guides in this series. However, there are a number of properties which need to be considered when choosing between different treatments. The following list, though not exhaustive, could be found useful. (It should be noted that not every property is required for every material.)

- resistance to entry of rain-water/water vapour

- resistance to deleterious gases (e.g. carbon dioxide in concrete)

- resistance to ultraviolet light (and therefore to chalking of any resin component of the treatment)

- resistance to entry of chloride or sulphate ions

- acid resistance

- permeability to water or salts escaping from the substrate

- elasticity

- crack bridging capability (live and passive)
- degree of preparation of substrate necessary
- adhesion to or penetration into the substrate
- chemical compatibility with substrate
- compatibility of physical characteristics with substrate
- ease of application
- toxicity/flammability
- ease of overcoating
- tolerance of substrate variability
- high tolerance of UK weather during application
- resistance to the growth of algae
- attractive appearance
- durability in relation to condensation, freeze—thaw or immersion cycles, application of salt, etc.
- length of maintenance-free life
- resistance to removal of graffiti by organic solvents
- relative economy of total treatment process
- proven use

Materials should preferably be non-toxic, easy to mix and apply, insensitive to rain and moisture during application and cure, and be tolerant of severe drying conditions, dry substrates, high temperatures, strong winds and poor surface preparation.

7.13. Cleaning

Aggressive cleaning of concrete or masonry can decrease durability considerably, particularly where grit blasting and acids or strong alkalis are used. Water soaking followed by high pressure water jet cleaning is less harmful. Stiff bristle, nylon or non-ferrous wire brushes can be used for localized difficult areas. If this method proves inadequate then carefully controlled wet grit blasting may be employed. Such treatment should be entrusted to a specialist company employing operatives fully experienced in nozzles, pressures and grits to achieve adequate cleaning without damage. A biocidal treatment should follow this to prevent algal growth. Full advice is given in BS6270 (see bibliography at the end of this chapter).

Select bibliography

1. Ashurst J. *Mortars, plasters and renders in conservation*. Ecclesiastical Architects and Surveyors' Association, London, 1983.
2. Ashurst J. and Ashurst N. *Practical building conservation*: Vol. 1, *Stone masonry*; Vol. 2, *Brick, mortars, earth*; Vol. 3, *Plasters, mortars, renders*; Vol. 4, *Metals*; Vol. 5, *Wood, glass, resins and technical bibliography*; Gower Technical Press, London, 1988.
3. British Standards Institution. *Cleaning and surface repairs of buildings*: Part 1, *Natural stone, cast stone, clay and calcium silicate brick masonry*, 1982; Part 2, *Concrete and precast concrete masonry*, 1985; Part 3, *Metals (cleaning only)*, 1991. BS6270, BSI, London.
4. British Standards Institution. *Code of practice for falsework*. BS5975, BSI, London, 1982.
5. Building Research Establishment. *Repair and renovation of flood damaged buildings*. BRE Digest 152, BRE, Garston, 1973.
6. Building Research Establishment. *Drying out buildings*. BRE Digest 163, BRE, Garston, 1974.
7. Building Research Establishment. *Principles of joint design*. BRE Design 137, BRE, Garston, 1977.
8. Building Research Establishment. *Control of lichens, moulds and similar growths*. BRE Digest 139, BRE, Garston, 1977.
9. Building Research Establishment. *Painting walls: Choice of paint*, BRE Digest 197; *Faults and remedies*, BRE Digest 198, BRE, London, 1977.
10. Building Research Establishment. *Site use of adhesives*. BRE Digests 211 and 212, BRE, Garston, 1978.
11. Building Research Establishment. *Cleaning external surfaces of buildings*. BRE Digest 280, BRE, Garston, 1983.
12. Building Research Establishment. *Building mortars*. BRE Digest 362, BRE, Garston, 1991.
13. Construction Industry Research and Information Association. *Structural renovation of traditional buildings*. Report 111, CIRIA, London, 1986.

14. Historic Buildings and Monuments Commission. *Addendum to Property Services Agency Schedule of Rates for Building Works.* HBMC, London, revised annually.
15. Institution of Structural Engineers (Scottish Branch). *Proceedings of Symposium on the structural repair of Scottish tenements.* IStructE, London, 1984.
16. Lambert B. S. *Remedying defects in older buildings.* Technical Paper 89, Chartered Institute of Building, London, 1988.
17. Property Services Agency. *Schedule of Rates for Building Works.* PSA, London, revised annually.
18. Rainger P. *Mitchell's movement control in the fabric of buildings.* Batsford, London, 1983.
19. Richardson B. A. *Remedial treatment of buildings.* Construction Press, London, 1980.
20. Royal institute of Chartered Surveyors. *BMI Special Report No. 167 — Condition surveys.* RICS, London, 1988.

8

Legal aspects

8.1. Engineer's responsibilities

In meeting his obligations there are a number of important considerations to be borne in mind by the investigating/reporting engineer. The engineer who prepares a report does so in the knowledge that considerable reliance will be placed on his findings. He should also be aware that in addition to his contractual duty to the client who has commissioned the report there is a great likelihood that other (non-contracted) parties will at some time have a sight of the report and will also place their reliance on it.

It follows that when carrying out investigations for any purpose (including particularly preparing a report) a duty of care is directly owed to the public at large (i.e. in law, his 'neighbour') as well as any directly interested parties. A simple example is an investigation that reveals a structure is unsafe to the point of collapse into the adjoining street. While the owner (assuming he is the client) must be informed, so must other parties involved in public health and safety. Under the current building regulations there are clearly set out, in the Building Act 1984, Sections 76 and 77, special procedures for dealing with such cases and in some emergency situations the local police should also be advised so that they can control or prevent public access into the dangerous area involved. Hopefully it should be possible to persuade the client of the need for such action, but the engineer must make his own decisions and act upon them. Another example of an unforeseen outcome would be the case if, during an investigation to establish structural strength, a serious fire risk is uncovered, although not within the original terms of reference. Some further action is then clearly called for on the part of the engineer.

In the proper performance of his duties the engineer is likely to visit construction sites or buildings and structures where work is being carried out. Although it is unlikely that the engineer will have received the specific training that would render him a competent person within the meaning and definitions of the Health & Safety at Work Act 1974, by the very nature of his normal training and experience he will be much more likely than the 'man in the street' to be aware of any improper practices followed by those working on the site. If, therefore, his suspicions are aroused that any such improper practices could lead to a danger to the health and/or safety of anyone, then he should take steps to make his views known to someone in authority on the site; failing this he should inform the workers themselves. It should be noted that there is a government inspectorate which deals with these matters and the local department can be found upon application to the local authority for the area involved.

It is also possible that in following an investigation an engineer may come across some feature or phenomenon which could give rise to an environmental health danger, and while his general experience and training is unlikely to enable him to deal fully with such a problem, he would be required to take appropriate action to obtain specialist advice and/or assistance. In this connection it should be noted that local authorities have environmental health sections which can provide assistance and/or send out environmental health officers to deal with any suspected contraventions of the regulations governing such matters.

The specific area of fire risk is one where assistance can be readily obtained from locally acting fire officers. These officers also act as a body responsible for issuing fire certificates on many classes of buildings which could be withdrawn if a hitherto undetected aspect of fire risk is uncovered. It should be noted in passing that matters relating to fire come under the administrative control of the Home Office, while other building matters (including administration of the Building Regulations) are covered generally by the Department of the Environment.

Another area of potential legal difficulty is where the engineer's investigations indicate a seriously unsatisfactory state of affairs but his recommendations for further action are ignored by the client/employer. Here he should make it abundantly clear to his client what his views are by means of a written document (sent recorded delivery if necessary) setting out the likely consequences of failure to heed his recommendations. If this approach does not give the desired result, the engineer may well then need to consider taking legal advice on how to proceed.

Failure to exercise reasonable skill, care and diligence would clearly leave an engineer open to claims for negligence which, if he is an employee (under current master and servant law), leads to his employers being liable for his activities. Risk of claims under such circumstances are described as 'vicarious liability'. Self-employed (partner/director), engineers naturally attract any such liability directly upon themselves (and also their fellows according to the conditions of their partnership or incorporation).

It should always be borne in mind that in reporting, the engineer is giving evidence, albeit in an informal manner. The logical extension of this position is to consider that any such reporting could become involved in legal action where it would be included as evidence or the engineer could be called as a formal witness to give direct evidence to the court. When investigations lead to court actions the engineer involved would invariably be described as an expert witness and Reynolds and King (12 in bibliography) give excellent guidance on this particular aspect.

8.2. Scope of work

Whereas a few reports are based on desk-top study only, the vast majority involve some form of survey of existing buildings. It is in this area where considerable risks exist which must be faced by the engineer in a comprehensive and proper manner before any work is undertaken. In this connection a general condition survey and report can involve costs of the order of, say, 0.1% of the building in question. Thus for a relatively small sum the whole responsibility for a building worth, for instance, one thousand times the fee involved can be taken by the investigating engineer. Conditions limiting liability can often be negotiated in such cases, and it is worthwhile establishing these conditions at the outset.

8.3. Terms of appointment

The outcome of the *Sutcliffe* v *Sayer* case (1987) depended on the 'terms of reference'. 'The terms of reference and conditions of engagement must be established unequivocally and in writing at the outset. It is important to obtain a clear mandate on the scope and extent of the work involved and if the engineer at any time forms the view that his actual commission does not cover the area of potential risk then this he must make clear and seek further instructions'.

It is normal to work in accordance with the terms and conditions of the Association of Consulting Engineers. Other conditions may well imply unduly onerous obligations. A serious consequence of departing from the normally accepted scope of conditions could be the negation of professional indemnity insurance cover which, contrary to popular misconception, does *not* cover every risk; if there is doubt then this must be cleared up with insurers or their brokers before finalizing any conditions.

8.4. Professional conduct

An engineer working on an investigation should observe rigidly the rules and conventions of professional conduct. Most professional organizations lay down such rules and an investigating engineer should be scrupulous in informing any other professional whom he is aware is likely to be involved of his interest and intentions. This is normally a matter which should be resolved when negotiating before undertaking a commission. In the circumstance where the commissioning client withholds his approval to any such observance, then the investigating engineer should consider obtaining advice from the professional body to which he belongs. (Both the Institution of Civil Engineers and the Institution of Structural Engineers have published rules of conduct covering these matters and the Association of Consulting Engineers also bind their members to similar conditions).

8.5. Experience

Engineers should only undertake work that they are qualified and experienced to do, as there is considerable risk in working outside recognized expertise.

8.6. Unseen items

Engineers should not comment on unseen items without qualification. For example, where there is no evidence of related distress and, as foundations are very difficult to inspect and check directly, it is permissible to state that 'there is no evidence of foundation behaviour that can be taken to suggest that these have not been designed and constructed in a normally acceptable manner'. Only in these circumstances is there no call for further investigation. If, however, there is some evidence of undue foundation movement then this must be commented upon and suggestions for further investigative action put forward if considered necessary. Failure to resolve this state of affairs could leave the engineer vulnerable.

Another pitfall relates to areas which are difficult to inspect because of access. It is not sufficient for an engineer to ignore a potentially important part of a survey or investigation for such reasons and some means should be devised of overcoming this problem.

8.7. Sampling

It is sometimes necessary to assess many items by examining only a few. For example, precast concrete window mullions are a frequent cause of complaint, and for various reasons, it is usually impracticable to consider inspecting every mullion in an initial investigation. The basis of reporting therefore must be made clear, e.g. 'On the assumption that the five/ten/twenty cases examined are reasonably representative of the whole number involved then conclusion A, B or C can be reached'. Clearly if one item in one hundred is satisfactory, it is obviously not possible to state that all the other ninety-nine are completely satisfactory and some statistical concepts must be addressed.

8.8. Recording

Full and accurate recording is essential. It is preferable to keep all site notes and dimensions in book form (rather than loose leaf) with the date, time and weather conditions fully recorded. Similarly if photographs are taken, whether used in the report or not, they should be located and identified so that future reference is possible. A key plan showing photograph positions should therefore be made as part of the site notes. It has been found of value to use a miniature tape recorder while taking photographs so that a verbal description of every shot is available. Also of great value is a date and time device on a camera, if the first and last photographs on each film (or run) show the date and intermediate ones the time, then the order of photography can be established without going through the negative numbers. Again the negatives should be carefully stored so that further prints can always be obtained. For various reasons it is essential that photographs are logged and marked up as soon as the prints become available; failure to do this could result in the required feature not having been recorded clearly, or even more likely, great difficulty in making a proper identification some months afterwards. When using instruments and equipment, the site notes should refer to them e.g. 'levels checked with 3ft builders' spirit level'.

8.9. Independent assessment

It is important not to be 'led' or to seek evidence only to substantiate a client's case. The reporting engineer should therefore start with a completely open mind and only come to his findings through a process of elimination. In many cases a preliminary assessment of damage has already been made, but it is important to look at other aspects and locations as well as those of the previously reported damage so as to be satisfied as to the precise nature of the overall behaviour pattern when reporting finally.

8.10. Terminology

Terminology is very important. The use of the expression 'structural survey' should only be related to that work commonly carried out by a surveyor which deals with the whole of a building, including services, drains, finishes and other non-structural items. An engineering survey can be called a structural inspection but it is preferable to entitle a report with a specific, accurate and appropriate title, e.g. 'Report on suspected subsidence at 20−24 Cherry Road, Cambridge'. Another word to be used with caution is 'maintain'. In lease agreements the meaning is very severe and extensive and usually imparts a contractual obligation to hand back an item in the original condition at the expiry of the lease period without any allowance being made for fair wear and tear concepts or natural ageing and deterioration.

8.11. Assessing responsibility for defects

An engineer may be specifically instructed to carry out an investigation so that responsibility for a defect can be allocated and which invariably involves alleged negligence. This is an area where great caution and discretion is called for. In general terms a ruling from a court of law is necessary to establish that some party has been guilty of negligence and the reporting engineer is therefore well advised to avoid making any such direct allegations. In any event he will normally not have access to all the relevant facts and circumstances involved so that any conclusion could be premature. It is, however, fully permissible to follow the indications of the evidence at his disposal. For example, he could conclude that 'the excessive deflexion of the second floor slab does not appear to be related to unsatisfactory constructional practice and is unlikely therefore to be indicative of any insufficiency in the design process for which Messrs A. B. Cee & Partners were the consulting structural engineers'.

8.12. Further reading

There is a great deal of published work available which has a bearing on legal issues, the study of which could be helpful in many instances. While it is not possible to provide a complete catalogue, a short list is given in the bibliography at the end of this chapter. In addition the different legal actions in the courts that have made case law under English Common Law (see below) are all reported in various official law reports, but other publications carry regular reports of such cases as they occur and accurate reviews are contained, for example, in the journals *Structural Survey* published quarterly by HS Publications and *Building Technical File* published quarterly by Building (Publishers) Ltd. The following lists are given to assist in providing some further background to the legal position with regard to construction of buildings. The first sets out the main relevant Acts of Parliament and statutes (there are of course many other statutes which can bear on building matters). More comprehensive lists are given by Wilkie and Howells, and by Owen (11 and 14 in bibliography).

- Statutes and Acts of Parliament

  o The Public Health Act 1936
  o The Public Health Act 1961
  o The Building Act 1984

(The Building Regulations 1985 are based upon the above three Acts)

  o The Health & Safety At Work Act 1974

(The Factories Act 1961 and the Office Shops and Railway Premises Act 1963 are generally consolidated in the 1974 Act)

  o The Defective Premises Act 1972
  o The Housing Act 1961
  o The Housing Act 1964
  o The Repair of Boniface Buildings Measures 1972 (deals with ecclesiastical premises)
  o The Sports Ground Act 1975
  o The Latent Damage Act 1986

The next sets out the more recent court cases where decisions reached (and sometimes these were only finalized at the appeal stage) have had an important bearing on subsequent litigation and as such make salient points in the development of common law as applied to building and construction.

- Reports of court actions

  o *Anstruther & Others* v *McOscar* (1924)
  o *Jacobs* v *London CC* (1930)
  o *Donaghue* v *Stevenson* (1932) (the snail in the bottle case)
  o *Bolam* v *Friern Hospital* (1957)
  o *Cartledge* v *Jopling* (1963)
  o *Clay* v *Clump* (1963)
  o *Hedley Byrne* v *Heller & Partners* (1963/1964)
  o *Moresk Cleaners* v *Hicks* (1966)

- *Driver* v *William Willett* (1969)
- *Warboys* v *Acme Investments* (1969)
- *Home Office* v *Dorset Yacht Co.* (1970)
- *Dutton* v *Bognor Regis* (1972)
- Sutcliffe v *Thackrah* (1974)
- *Greaves Contractors* v *Bayham Meikle* (1975)
- *Kensington AHA* v *Wettern Composites* (1976)
- *Oldschool* v *Gleeson* (1976)
- *Sparhan Souter* v *Town & Country* (1976)
- *Anne* v *Merton* (1978)
- *Batty* v *Metropolitan Property* (1978)
- *Mount Albert BC* v *Johnson* (1979)
- *R.V. Cardiff CC Ex Parte Cross* (1981)
- *Dennis* v *Charnwood BC* (1982)
- *Junior Books* v *Vetechi* (1982)
- *Acrecrest Ltd* v *W S Hattrell & Partners* (1983)
- *Pirelli* v *Oscar Faber* (1983)
- *Peabody* v *Sir Lindsey Parkinson* (1984)
- *Davy Chiesman* v *Davy Chiesman* (1984)
- *Geo. Wimpy* v *D V Poole & Others* (1984)
- *Wainwright* v *Leeds CC* (1984)
- *Investors in Industry* v *South Bedford DC* (1985)
- *Parker* v *Camden LBC* (1985)
- *Ketterman* v *Mansel Properties* (1985)
- *Quick* v *Taff-Ely BC* (1985)
- *Equitable Debentor Assets* v *William Moss* (1986/1987)
- *Sutcliffe* v *Sayer* (1987)
- *Rickards* v *Kerrier DC* (1987) (first case on the Building Regulations 1985)
- *D & F Estates* v *Church Commissioner for England* (1989)
- *Murphey* v *Brentwood District Council* (1990)
- *Department of the Environment* v *Thomas Bates* (1990)

Again it must be accepted that while the cases cited are generally recognized as being significant there are many others also of relevance. There are also other recent important cases which have not yet been fully reported. Most of the cases listed plus many others are referred to with details of judgements given in the bibliography,(11, 12 and 14) the later ones being reported in journals.

It should be noted that the law of tort has recently seen a reversal from the position established in *Anne* v *Merton* in 1978, but it is still prudent for an engineer to consider the possibility that his actions could be the subject of a tortuous action notwithstanding the outcome of recent cases in the courts.

It should be appreciated that with the advent of the common market for the European Community many aspects of law related to engineers' work will undoubtedly change and it will be necessary to follow such developments as they are codified and rectified.

Select bibliography

1. Akroyd T. The engineer at risk — a personal view of professional responsibility and liability. *Proceedings of Institution of Civil Engineers Conference on Conservation of engineering structures.* Thomas Telford, London, 1989.
2. Atkins Planning. *Latent defects in buildings.* Atkins Planning, 1985.
3. British Standards Institution. *Building maintenance management* BS8210, BSI, London, 1986.
4. Butler R. *Latent damage.* Lloyds of London, 1987.
5. Carper K. L. (ed). *Forensic engineering — learning from failures.* American Society of Civil Engineers, London, 1986.
6. Greenspan H. F. *et al. Guidelines for failure investigation.* American Society of Civil Engineers, New York, 1989.
7. International Association for Bridge and Structural Engineering and Institution of Structural Engineers. *Liability.* IStructE, London, 1985.
8. Institution of Civil Engineers. *The presentation of engineering evidence.* ICE, London, 1946.
9. Institution of Structural Engineers. *Rules of Conduct. Guidance Note 3, Checking and appraisals; Guidance Note 4, Surveys of residential properties.* IStructE, London, 1986.
10. Legal Studies and Services Ltd. *Latent Damage Act 1986.* Symposium papers. Legal Studies and Services Ltd, London, 1986.
11. Owen S. *Law for the builder.* Longman, London, 1987.
12. Reynolds M. P. and King P. S. D. *The expert witness and his evidence.* BSP Professional Books, Oxford, 1988.
13. Royal Institution of Chartered Surveyors. *Structural surveys.* RICB, London, 1985.
14. Wilkie M. and Howells R. *Practical building law.* Mitchell, London, 1987.

9

Methods of access

9.1. Health and safety

The most significant change to have occurred in recent years in relation to means of access is the Health and Safety at Work Act 1974. The Act makes every employer responsible for the health, safety and welfare of all his or her employees when they are at work and requires the preparation and regular updating of a written policy statement declaring how this responsibility will be undertaken. Specific requirements relate to the provision of plant and the system of work which must, as far as reasonably practicable, be safe and without risks to health. Particular responsibility rests with the employer with regard to the use, handling and transport of articles and substances and to the provision of instruction, training and supervision. Moreover the Act requires access to and egress from a place of work to be provided and maintained, as far as reasonably practicable, so that it is safe and without risks to the health, safety and welfare of employees.

Many appraising engineers will be employees of one firm but will be working on a site where most of the workforce will be employees of another. It is important therefore that the appraising engineer should discuss any proposals for access to or within the site with the safety management of the principal employer, otherwise the responsibilities of the latter could well be compromised by the actions of the employees of the former. It follows that the principal employer should be kept informed of the daily (perhaps, where appropriate, hourly) location of the appraising engineer's staff.

Three common-sense rules apply to access

- Always keep someone within contact — voice contact with someone else is normally satisfactory; someone must be in earshot for a call for help to be of any use. However, in some situations where voice contact is not easily possible, a more sophisticated means of contact is called for. This rule applies not only to individuals entering rooms on their own but also to groups of people entering places where they are likely to be unseen by others for more than a few minutes.

- Always keep means of escape clear — they may be needed when there is no time available to move things out of the way.

- Never use a structure for access unless there are good *positive* reasons for believing that it and its supports are strong and rigid enough for the purpose. In certain circumstances this may require a very thorough inspection of ladders, platforms and the like before use, in others — for example, where the principal employer has his own employees working on a structure — this may not be so necessary.

9.2. External inspection

For property of a domestic scale it will usually be possible to gain access with either ladders or scaffold towers. Roof ladders must always be used on sloping roofs and crawling boards on fragile ones. Access to many larger buildings can often be gained by using a window cleaner's cradle or by having a similar but temporary device provided. A useful piece of equipment is the mobile 'cherry-picker' type of vehicle which has the additional advantage of being able to take the inspecting team below the level of the structure or ground upon which the vehicle is sitting. Such vehicles, for reasons of safety, must only be operated by trained personnel. For even taller buildings, or for a façade inaccessible to other means, abseilers with some knowledge of engineering and buildings can be hired.

When samples are to be obtained by drilling the methods outlined above are unlikely to prove entirely satisfactory on their own but, except on a blank wall, it may be possible to anchor a cradle through the windows to give the rigidity necessary for drilling to proceed.

Scaffolding is expensive and should only be used where there is no practicable alternative or where its continued use during remedial works would absorb a large proportion of the cost. However, the scaffolding must, if continued use is proposed, be suitable for the more demanding use during the latter phase.

Buildings beside or over water present particular problems. Any boats from which access is to be afforded must not only be secured against movement but are likely to tilt when a load is moved.

The need to provide access for the purpose of reading monitoring equipment can be obviated if care is taken to ensure that it can be seen clearly from a safe vantage point — perhaps using binoculars or through a nearby window.

9.3. Internal inspection

Folding ladders will prove to be a valuable means of access when inspecting the inside of a building. Access to loft spaces can usually be made through existing trap doors, and crawling boards should be used if the ceiling joists are of doubtful strength.

Instances have occurred where the access to cellars has been boarded over and even the tenants have been unaware that a cellar existed. As the majority of nineteenth-century buildings — particularly those in towns — had cellars for the storage of coal, a search for one should always be made. This is particularly important as dampness and lack of ventilation make cellars a breeding ground for the fungi which attack timber.

Fitted carpets and other floor coverings make the disturbance of domestic ground floors unpopular. If it is considered important to view the underside of such a floor an alternative is to use one of the optical probes described in Chapter 10, together with a strong source of light, through the openings of an existing air-brick.

9.4. Below-ground access

Trial holes must be adequately shored and the safety of third parties considered. Strong fencing may be necessary around a hole even if only intruders would be injured by falling into it.

Deep holes, either new or existing (e.g. wells) should be entered with caution. Breathing apparatus must be available and, if not being used by the inspecting engineer, his assistant on the surface must have it immediately to hand. The presence of explosive gas must always be considered, and any risk of fire generated by sparks or matches carefully considered.

Similar precautions apply to the inspection of sewers, and in addition the engineer should be fully protected from contact with infectious material. Medical advice should be sought as to the advisability of inspection and/or inoculation.

9.5. Underwater access

Diving gear should only be used by qualified divers who are up-to-date with their regular qualification dives. (NB. Divers have to spend a minimum period of time below water every six months to remain qualified.)

Information regarding the type of instrumentation for use under water is given in Chapter 10.

In the clouded water which is likely to be normal around most structures to be inspected, the only way to inspect an underwater structure thoroughly is to isolate it by means of of a cofferdam and pump out the water. However, this should be done with caution, as the change in water regime may cause distress in retained soils by the alteration of pore water pressure.

Select bibliography

1. Construction Industry Training Board, *A guide to practical scaffolding.* CITB, King's Lynn, 1987.
2. Health and Safety at Work etc. Act 1974, HMSO, London.
3. Institution of Structural Engineers. *Guide to surveys and inspections of buildings and similar structures (includes as Appendix A.* Abstract from the Health and Safety at Work etc. Act 1974, IStructE, London, 1991.

10

Instrumentation

10.1. Introduction

This chapter reviews the instrumentation available to assist the engineer in carrying out the investigation of a building structure but does not deal with instrumentation used for monitoring which is the subject of Chapter 11, nor with instrumentation relating to the investigation of specific materials for which the reader is referred to the appropriate guide in this series.

10.2. Visual inspection

A range of devices exists to assist visual examination and recording. In recent years there have been developments in instrumentation which enable cavities and other inaccessible areas to be examined from a remote position.

Cameras

Normal 35 mm single lens reflex (SLR) cameras are reasonably compact yet allow a variety of interchangeable lenses to be used. A 28 mm wide angle lens allows the camera to be positioned closer to the subject than the more usual 50 mm lens. Wider angles than that of a 28 mm lens lead to distortion. Telescopic lenses (e.g. 70–210 mm) allow distant objects to be photographed and can often also be used as macro-lenses for close-up work. Automatic exposure is essential unless a light meter is used. Polaroid cameras allow results to be seen before leaving the site but duplication of results is poor. Shutter speeds slower than one-thirtieth of a second require the use of a tripod, monopod or other support. A polaroid filter (PL1) can be used to reduce the effects of glare resulting from the reflection of light from glass or water surfaces and other filters may be used to enhance particular features.

Lighting

Care should be taken in positioning light sources for use with photography. If the light source is placed behind the camera very little detail will show up. If it is to one side, shadows will be created which can indicate discontinuities, ridges, etc. In some cases it will be desirable to take several photographs of a subject with the light source in various positions. Whatever form of light is used the aperture required by an automatic exposure device or a light meter should be increased by one stop for side lighting and two stops when the light source is behind the subject.

Photographic film

The popular makes of both negative and reversal (transparency) film at 100–200 ASA are cheap and give good results under normal lighting conditions using a hand-held camera. Faster films are more expensive and less satisfactory as detail is lost due to the more grainy nature of the results. The use of a tripod or other support to allow longer exposures or the use of artificial lighting are better alternatives. Slow films (e.g. 25 ASA) give better results but are rarely suitable for use with hand-held cameras in natural lighting. Finer grain films are available and a specialist supplier should be consulted regarding their suitability for a particular application. Transparencies have several advantages over negative film but require more precision when deciding on the exposure. However, transparencies can be created from good prints if required. Colour is preferable to monochrome but printing in colour in reports is expensive.

Magnification devices

Telescopes and binoculars are useful for detailed examination of parts of a structure that are visible but not easily reached. Although a high magnification is required for detailed inspection, it can be difficult to target high magnification binoculars — models provided with variable magnification can overcome this problem. As well as the familiar fixed focus magnifying glass there are torch-light magnifiers which are obviously useful in darker situations, and measuring magnifiers which enable a more precise description of what is seen. For really close-up work pocket microscopes are available and there are also specialist microscopes which, for example, allow measurements to be taken both across and within the plane of view.

Optical probes
The fibrescope consists of two bundles of fibre optics in a flexible protective sheath and is useful for inspection within inaccessible cavities. One of the fibre bundles carries the light source to illuminate the object to be viewed and the other carries an image of the object to the eye of the observer. Some fibrescopes have a controllable articulated section near the tip which enables the observer to obtain views in several directions. Accessories which can be used with a fibrescope include 35 mm SLR cameras, closed-circuit television cameras and video recorders.

The borescope is also used for inspecting cavities but differs from the fibrescope in that it is rigid and the optical viewing system consists of a series of lenses. Borescopes are cheaper than fibrescopes and images are much clearer. A variety of objective heads are available including wide-angle fish-eye and 90° side-viewing heads, the latter being useful when investigating faults in cavity walls. Borescopes can use a similar range of accessories to fibrescopes.

Closed-circuit television
Long cavities, such as occur in chimneys or sewers, can be inspected using a closed-circuit television camera and cable, winch, lighting, control unit and monitor. Video tape recordings can be taken of part or all of the survey and can include a digital display of camera position. Some systems enable still colour photographs of high definition to be taken in addition. Standard equipment can be used in cavities of 225 mm internal diameter or larger and less sophisticated equipment in cavities from 100 mm diameter upwards. There are, however, even smaller cameras available (down to 17.3 mm) for specialist applications.

Aerial surveys

Conventional aircraft and helicopters are expensive and may have to be flown too far from the object to be inspected to be of much use. Experiments have been undertaken with cameras mounted in radio-controlled model aircraft, on kites and more recently, on radio-controlled model helicopters. Good results have been obtained using skilled operators.

Underwater inspection

Some divers are qualified engineers and such should be used for significant underwater inspections. When an engineer diver is not available an alternative is for the diver to manipulate an underwater camera while directed through an intercommunication system by an engineer watching the camera output on a monitor above the surface. Video systems used under water do not usually provide enough fine detail or give sufficiently good resolution or colour quality. Still photography, however, can give high resolution if a 70 mm camera is used and if reasonably accurate colour reproduction can be obtained.

Photogrammetry

When drawings of a building elevation are unavailable, particularly where complex architectural details are incorporated, the use of photogrammetric techniques can considerably reduce the time necessary to carry out a detailed survey. A clear view of the façade to be surveyed is necessary however, and the work can only be carried out by surveyors specializing in this technique.

10.3. Measurement

It is assumed that the engineer is familiar with normal survey practice and the use of theodolites and other surveying instruments in common use.

Plumb

The traditional plumb bob is satisfactory for heights of a few metres in still conditions providing the bob itself is sufficiently heavy and is suspended in water to dampen oscillations. For greater heights an accurate theodolite or an autoplumb should be considered. The latter can be read down to better than 0.5 mm in 10 m. The instrument is used at ground level and targeted on a reference point firmly cantilevered from the structure being surveyed. When accuracy is not a critical requirement a 900 mm spirit level may be used.

Cracks
The width of cracks can most easily be measured either by the use of a card on the edge of which different widths of cracks are drawn out accurately, or by using callipers. Without such help there is a tendency to overestimate the width of cracks. There are now several different types of calliper available, the most convenient of which use electronics to provide a digital reading. A depth gauge is useful for measuring fractures in the corners of walls. It should not be forgotten that displacement — both horizontal and vertical — as well as the width of cracks, may be of importance and a steel straight-edge and steel rule are at present the only suitable tools for this purpose.

10.4. Location of hidden detail

Breaking-out can sometimes be avoided by using instruments to detect hidden items. For example, cover-meters are useful for finding ferrous materials hidden in walls etc. Although X-rays can be used they are expensive, involve bulky equipment and are hazardous. An alternative for external walls where the internal temperature is higher than the external, is to use infra-red scanning or photographic techniques which will show up significant hidden details if these affect the flow of heat through the wall. However, care should be taken to note the position of heat emitting devices such as radiators before analysing the data so produced. Rendering of external walls should be broken away if there is any suspicion that it is hiding defects. The skilful use of radar may reveal voids, metals and changes of density.

10.5. Stress and strain measurement

Strain gauge techniques
A material's electrical resistance or conductivity changes with a change in length under stress and such changes can be measured using strain gauges. However, skill is needed in their use if

spurious results are to be avoided. Useful tables showing the most appropriate type of strain gauge for selected applications and the types of adhesives which can be used for mounting them are given by Collacott (2 in bibliography).

Ultrasonics

Because of their good elastic properties most metals readily transmit ultrasonic vibrations and these are likely to be scattered or reflected by flaws. Technology which makes use of this phenomenon has been well developed in the inspection of welds. Similarly, long members of regular cross-section using other materials (e.g. concrete piles) have been successfully tested using ultrasonics.

Resonance testing

Structures may be vibrated both to locate defects and to determine structural characteristics. However, the technique is not applicable to squat structures or members. Vibration is induced either by applying a known external excitor, for example through the use of a motor with eccentric masses, or by applying or releasing previously applied load suddenly. In order to find structural faults the normal response of the structure being tested needs to be calculated or, where a series of similar structures or members exist, significant differences between the response of individual members may be sufficient to indicate faults. Testing can be repeated when repair work has been carried out to confirm that it has been effective.

Where there is doubt about the ability of a structure to stand up to large vibrations which it is likely to meet in practice, dynamic testing on a smaller scale can be used to predict behaviour under higher loads.

10.6. Dampness

A number of electronic meters with probes are available with which the relative dampness within the surface layers of a structural member can be measured.

Select bibliography

1. Building Research Establishment. *Simple measuring and monitoring of movement in low-rise buildings: Cracks*, BRE Digest 343; *Settlement, heave and out-of-plumb*, BRE Digest 344. BRE, Garston, 1989.
2. Collacott R. A. *Structural integrity monitoring*. Chapman and Hall, London, 1985.
3. Laurens D. Thermographie infrarouge applique a des batiments anciens [Application of infra-red thermography to old buildings]. *International Association for Bridge and Structural Engineering Symposium Proceedings*, Venice, vol. 46, pp. 117–127, 1983.
4. Uren J. and Robertson G. C. Recording cracks photographically. *Structural Survey*, 1987, **5**, No.4.

11

Monitoring

11.1. Introduction

Monitoring techniques are used to measure the response of a structure to changing conditions, to seek evidence of cyclic elements in crack behaviour, or to assess the rate at which deterioration may be taking place. Additionally, sophisticated monitoring systems may be employed to give warning of overload or potential collapse. Monitoring of ground and groundwater movements is dealt with in the *Appraisal and repair of foundations* in this series.

11.2. Monitoring structural movement

Accurate diagnosis of a structural problem frequently requires the measurement of structural movement in relation to time, to diurnal or seasonal changes in temperature or humidity, or to the application of a load. Non-reversible movement at a reasonably constant rate can be measured by periodically re-surveying using the techniques mentioned in Chapter 10. Accuracy, however, requires that the locations from which measurements are taken are unequivocally and durably marked by using suitably inscribed non-ferrous metal or stainless steel targets secured to the structure. Accuracy will also be enhanced if the same instruments are used each time a measurement is taken. Photography can be quite useful in measuring change over time, particularly when transparencies of the same subject taken at different times are superimposed. It is of course important that the camera position, lens size, etc. are the same on each occasion.

11.3. Displacement across structural cracks

The Demec strain gauge is a useful tool for monitoring movement across cracks in two directions. Two location points are provided on one side of the crack and one on the other. These points can consist of stainless steel discs of about 6 mm diameter, each having a small hole drilled at the centre for accurate positioning of the points of the gauge, which are fixed to the structure with a suitable adhesive. In very exposed locations a more durable form of point firmly set within a hole in the structure may be preferable. On each occasion when a measurement is to be taken, the dimensions from the single point to each of the two points on the other side of the crack are taken and from these, by trigonometry, both the lateral and longitudinal displacements can be calculated. The locating points in a system of this kind need not be unsightly and an accuracy of 0.025 mm can be achieved. A steel straight-edge and rule can be used to monitor movement in the third direction. Glass tell-tales should never be used in monitoring (once broken they are useless and breakage may be caused by factors other than structural movement).

11.4. Diurnal and seasonal movement

The costs of the sophisticated equipment necessary to monitor structures continuously is likely to preclude its use except in exceptional circumstances or for experimental purposes. Periodic re-surveys are therefore more common. The vagaries of the British climate are such that a single set of readings taken in the winter compared to a set of readings taken in the summer are unlikely to be of much value. For seasonal variations, readings should be taken at monthly intervals throughout the year and additionally when excessively hot or cold conditions occur. For diurnal variations minimal intervals of 3 hours over 48 hour periods on at least four occasions spread throughout the year should prove satisfactorily accurate. It should be borne in mind that the diurnal variation on the south side of a building will differ considerably from that on the north. It is also important to record the air temperature both inside and outside the building simultaneously. Cyclic variations can lead to a 'ratchet effect' of expansion in one direction followed by contraction in the same direction. However, friction or foreign matter in cracks can prevent full contraction.

11.5. Short-term movement

The methods to be used to measure the response of a structure to short duration events, such as wind buffeting or the passage of vehicles, will depend upon the predictability of the events occurring. When the occurrence of an event can be predicted or an event can be caused to occur at a pre-determined time, readings on instruments set up for the event can be taken by an observer. In less predictable situations data-logging techniques will be necessary. Such techniques are constantly developing but some have been used successfully for several years on buildings such as St Paul's Cathedral. Advice on their application and use should be sought from experienced sources such as the Building Research Establishment.

11.6. Stress and strain monitoring

The principle for monitoring forces in a structural member is to attach to or embed within the member a transducer which transmits readings to a data logger or computing device. British Rail have developed a cylindrical transducer 29 mm in diameter and 28 mm deep which can be installed in a previously drilled and reamed hole on the neutral axis of rails which enables the force in the rail to be measured. Devices such as this can only measure the change in stress from that which prevailed at the time of their installation. The absolute value of load or stress may be more difficult to determine.

11.7. Vibration monitoring

Monitoring of vibrations in a structure requires specialized techniques and reference should be made to *Structural integrity monitoring* by R. A. Collacott (2 in bibliography).

11.8. Monitoring deterioration

Methods of monitoring the deterioration of specific materials are described in the appropriate guides in this series. Coloured photographs can play an important role in plotting the progress of deterioration, particularly if each photograph is indexed and annotated to give the date, the time of day, the position from which it was taken, and other relevant information. The same make and type of film should be used on each occasion.

11.9. General requirements for a monitoring system

In considering the type of monitoring system to be used for a particular application the following must be borne in mind.

- There should be minimum interference with normal operations

- The system should be robust and durable enough to withstand the effects of the weather it is likely to be subjected to and should be so placed as not to invite vandalism or interference by those not involved with the investigations

- Monitoring positions should be readily accessible or some form of remote reading (perhaps through a window) should be considered

- Although enough indicators are required to give a true picture of change, it should be remembered that the more indicators there are to read, and the less accessible they are, the more costly will be the monitoring programme. However, with skilful location of the indicators the extra costs may not be significant when compared with travelling, report writing and the potential loss of information by not having a back-up set of readings. For the same reason, the intervals between readings should be no shorter than is necessary for accuracy.

Select bibliography

1. Building Research Establishment. *Simple measuring and monitoring of movement in low-rise buildings: Cracks*, BRE Digest 343; *Settlement, heave and out-of-plumb*, BRE Digest 344. BRE, Garston, 1989.

2. Collacott R. A. *Structural integrity monitoring*. Chapman and Hall, London, 1985.
3. Hume I. *Monitoring of structures and landscape*. Historic Buildings and Monuments Commission, London, 1986.
4. Moore J. F. A. *Monitoring of building structures*. Blackie, Glasgow, 1991.

12

Load testing of structures

12.1. Introduction

The structural analysis of an existing structure is not an exact
science nor can analysis easily deal with three-dimensional effects
such as beams stiffened by floors, columns strengthened by infill
panels, load sharing and alternate load paths. Moreover, while
much can be learnt about the structure from examination of
existing drawings, site surveys and other detection work, there
may still be some uncertainty over the strength of the structure.
The engineer, faced with this uncertainty, will often find it helpful
to make use of a load test to confirm the adequacy of the structure.
The engineer will need experience, judgement and discretion in
carrying out load testing, for while this chapter gives guidance,
it cannot give hard and fast rules. Because load testing can be
an expensive and time-consuming operation the engineer needs
to consider carefully why the test is necessary, the likely value
of the results to the investigation in hand, the appropriate type
of test and the test procedures to be adopted, before embarking
on a test programme.

Specialist operations such as laboratory testing of structural
models and materials and dynamic load testing are not dealt with
in this chapter. Nor is the guidance given here necessarily
applicable to bridges, pylons, oil rigs, large petrol tanks and similar
'non-building' structures. This chapter deals only with static load
tests on existing building structures.

12.2. Reasons for load tests

Tests are carried out to check that the structure as a whole and its components are capable of carrying loads up to the ultimate and serviceability limit states given in codes of practice and other standards. Measurement of deflexions and other displacements, and of recovery on removal of load, can provide reassurance on some serviceability limit states. However, only long-term tests can provide evidence of the effects of creep, shrinkage and swelling, fluctuation or reversal of load, temperature effects and the like. When, on some occasions, this information is necessary the engineer will have to research BRE digests, CIRIA reports, technical papers, etc. The overwhelming majority of static load tests fall into the following categories.

- Change of use — the client requires increased load or change of load, the effects of which on the structure cannot be proven by usual calculations.

- Damaged structures — fire damage, deterioration due to lack of maintenance or other causes, accidental damage, overloading or any similar cause resulting in suspicions as to the structural integrity of elements or of the building structure as a whole.

- Non-compliance with specification — when workmanship or materials do not comply with the specification or standards and when the structural design departs from codes of practice or building regulations, it may not always be necessary to condemn or demolish.

- Sample checks — sometimes routine checks for quality assurance schemes, sample checks on mass-produced elements and the like are necessary. During the scare regarding high alumina cement, thousands of prestressed precast concrete floor units were load tested.

- Strength and serviceability checks — some parts of a structure, or even the whole structure, cannot always be justified with confidence solely by calculations. To assess strength, stability and serviceability, load testing can enable deflexions, strains (and by deduction, stresses) and other movements or distortions to be measured. The engineer will determine the strain, deflexion, and other criteria for acceptance and then undertake load tests to check for compliance.

12.3. Types of tests

In the same manner in which the engineer will clarify and define the reason for the test, he should define the type of test to satisfy that reason. The types of test are as follows.

- Site acceptance tests — these tests check structural behaviour under enhanced design loads (often, for example, double the live load) to determine adequate safety factors. The test should also include measurement of the amount of recovery of deflexion (often 90% is recommended), etc. on removal of the test load. If the structure fails to recover or behaves poorly under test, the test should either be re-specified, or the structure strengthened or not accepted.

- Quality control tests — these are carried out on parts of the structure or structural elements to check compliance with the specification or acceptance criteria. Generally they are non-destructive tests but sometimes an element or joint is tested to destruction to determine the safety factors at ultimate load. Typical examples are timber joints, precast concrete floor units and pull-out tests on new types of reinforcing bars.

- Laboratory tests — it is sometimes possible to remove suspect elements and components and test them in a laboratory, under more controlled conditions than are possible on site. This can cost less and be quicker than site testing.

12.4. Test specification

Load test specifications are given in codes of practice for structural elements but not for complete interconnected structures. Design partial factors given in BS8100 can be employed to give guidance on structural adequacy. Other structures designed before the first issue of codes may be covered by building regulations, particularly useful are those of the former London County Council.

Static and dynamic loading
Although it takes time for an audience to fill a theatre, snow to

fall on to a roof or furniture to be installed in an office, such dynamic loads are treated by the designer following the application or suitable safety factors, as static. Some structures are sensitive to dynamic loads (e.g. soldiers break step when crossing steel bridges) since harmonic motion can be created, thus leading to an increased risk of collapse. However, the vast majority of building structures and their loading cases may be considered as static for testing and analysis.

Static behaviour

Loads which occur less frequently than once an hour and are applied for a duration exceeding one minute produce a static response if the load is applied at a reasonable uniform rate to avoid dynamic effects. Structures behave statically when their response to a given loading can be predicted, knowing the loading, the structural material, element sizes and structural form. Stiffness remains constant under load and deflexions and other movements are not amplified by the loading. When the bulk of a building load is static and the structure is robust the dynamic effect of loading (including wind loading) is usually negligible. However, when the bulk of the load is not static and the structure is relatively flimsy there can be problems of dynamic response.

Dynamic behaviour

Loads such as pile driving, operation of heavy forges and other regular impact loading which could cause the structure to oscillate or vibrate, and other forms of dynamic loading which could have a frequency near that of the natural frequency of the structure, may amplify the structure's response and in such cases they should not be treated as static loads for test purposes.

Sturdy structures (traditional, massively thick masonry walled dock warehouses, sturdy terraced housing) can have a fundamental frequency in excess of 100 Hz. Tall slender structures (pylons, skyscrapers, radio masts) may have frequencies as low as 0.25 Hz and large span suspension bridges even lower at 0.10 Hz. Parts of a generally stiff structure can exhibit dynamic response such as a long-span, lightweight roof trusses supporting flimsy sheeting not rigidly attached to the roof structure. On the other hand long slender cantilevers can vibrate significantly yet not be in danger of collapse if they are correctly designed and detailed.

Bedding-down loads

Some materials (particulary plain brick and blockwork, and concrete and timber to some extent) bed down — i.e. contract and do not recover on removal of load — under initial loading. This is not serious but such contraction, if recorded, unconsciously distorts the assessment of deflexion. Some types of structure, such as precast concrete columns supported on dry-pack, some glulam constructions using split-ring connectors, etc. also need to bed down. Bedding down settles the structure on its supports and releases friction, compression and other stresses. The magnitude of the bedding-down load should be sufficient to deform the structure (often 10% of the design ultimate load) but should not exceed 80% of the design, or proposed service load. The load should be applied (and removed) in five equal increments and the deformations measured at each increment. It may be necessary to repeat the loading process. The structure can be considered as bedded down when it has a + 10% recovery of deformation on removal of load.

Serviceability testing

Serviceability tests are fairly common when a change of use is proposed, damage or deterioration has occurred, or where remedial measures require checking for their effectiveness. The tests check a structure's ability to carry safely the full service load with acceptable deflexion or other deformation.

The serviceability test load should be applied in 5 to 10 equal increments (and each increment placed over a period of at least l minute to avoid impact stress) and removed in the same number of decrements. Measurements of deflexion, crack width, length, etc. should be taken at each interval of applying and removing the test load.

The test load is often taken as the lesser of either $1.25 \times$ the sum of imposed and dead load (existing or proposed) $+ 0.25 \times$ the dead load, or the maximum that the structure should be able to carry without permanent deformation. The test load should be distributed to include the design worst case or worst position. A test should be deemed satisfactory when

- the maximum deflexion does not exceed that given in a code or specification

- any existing cracking or deformation does not worsen in the test

- the structure shows no sign of distress after loading and unloading

If the structure does not satisfy these criteria then some limitation should be put on its use or consideration might be given to an alternative test.

Ultimate load testing
The common aims of such tests — in effect proof tests — are to prove that the structure will support the design ultimate load safely without damage and to determine its deformation and behaviour. The quantitative data collected will monitor the deformation and qualitative observation will determine the deformation behaviour and character.

The test is carried out when there is uncertainty about the load-carrying capacity of the structure, its possible excessive deflexion and deformation and the behaviour of supports, fixings and bracings. The test is also carried out when there is suspect workmanship, materials, construction, etc., and when elements such as precast units are of doubtful performance.

The dead and imposed loadings should be factored as specified in the relevant code and load combinations to give the worst case should be employed. Differing combinations may be necessary to give worst cases for bending, shear, torsion and direct compression or tension.

The test loads should be applied in 5 to 10 equal increments and sufficient deflexion measurements taken for both maximum deflexion and deflected shape. It is advisable to produce a load/deflexion graph during the test to ascertain whether or not the structure is behaving elastically. If the loading starts to cause non-linear behaviour the loading increments should be decreased and extra caution should be practised in loading. It can be advisable in such cases to remove the loads to check the structure's recovery and the magnitude of any permanent set. It will be obvious that the deflexions under ultimate load will be greater than under normal load and the engineer should estimate the highest likely deflexion before the test.

Any existing cracks or defects (or new cracks and defects under loading), should be carefully monitored, measured and photographed during the test. Typical defects which could affect the stiffness of the structure but which are in elements rarely designed as part of a structure are cracking of infill panels, breaking up of infill hollow tiles in floors and debonding of floor screeds, buckling of lateral stiffening, failure of movement joints, etc.

The test may be deemed satisfactory when

- the structure, or any part of it, does not fail under the worst combination of load

- deflexion and deformation adequately recover on removal of load and any residual deflexion does not affect the safety and function of any part of the structure

- any partial failure (cracking of screeds, infill panels, etc.) does not affect the overall stability or integrity of the structure to support ultimate load safely.

Testing to destruction
This type of test is mainly carried out on innovative prototypes in order to develop the innovation further. There is a limit to the value of calculations, model tests and laboratory investigation; often the only certain way to check theory and research is to load to destruction. This checks the predicted performance against the actual, determines the true factors of safety and checks the failure modes. Such tests can also be useful in evaluating the effects of accidental damage.

Loads should at first simulate the serviceability case, then the ultimate design load conditions and finally the failure load. Measurements and observations should be carried out at serviceability limit and ultimate design limit as described earlier, and particular care must be taken as the predicted failure load approaches. It is vital to note both the structural behaviour of critical parts of the structure and its overall performance.

12.5. *External influences on in situ tests*

A test in a laboratory can be carried out on an isolated element under controlled conditions of temperature, humidity, etc. with very accurate loading and sophisticated measurements. It is difficult, if not impossible, to achieve such precision on site. This is not necessarily serious if the influences are taken into account and can actually be beneficial in not blindly applying the artificial conditions of the laboratory to the 'real world' of the site.

The following are the main factors affecting site test results.

Load sharing in floors, roofs and walls

A loaded floor beam can transfer some load to adjoining beams by the action of the floor slab. Similarly a loaded roof truss can transfer some load to other trusses through purlins, sheeting and bracing. Loaded columns can disperse considerable load through in-built panels of brick or blockwork and slabs designed to span in one direction can, with distribution reinforcement and side continuity, exhibit partial two-way spanning action. Even isolated elements such as precast concrete floor units can disperse load to adjoining units through infill pots and topping screeds. Any element in a structure can therefore disperse load to adjoining structural elements and thus deflect and deform less than expected because it is carrying less than its intended load. To prevent this would entail isolating the element by cutting the alternative load path but this can be expensive, it can require making good after the test, and it is often unnecessary and impractical.

The practical method is to load an area of floor or roof of width at least equal to the span, and it is usually found that a member located on the centreline of the loaded area will be supporting the true test load. Alternatively the beam or truss can be loaded and the deflexion of adjoining structural elements measured so that the load sharing can be approximately estimated.

Temperature movements

Steel roof trusses, particularly with uninsulated roof covering, will be subject to stress and movement due to temperature differentials. Therefore it is advisable to note over a period of time before the test the behaviour (deflexion, movement, strain, etc.) of such trusses due to temperature changes. During the test the effect of any temperature change will be known and can be separated from the effects of the test load.

Moisture and time movements

Timber, bricks and concrete are subject to moisture movement and it may, on occasion, be necessary to be aware that the test loading results apply only to the moisture conditions at the time of the test. With time clay bricks expand and concrete shrinks and creeps, and allowance for this may be necessary in predicting the future behaviour of the structure.

12.6. Procedure before testing

A scaffolded and propped safety platform must be erected beneath beams, slabs and roofs in case of structural collapse. It should be designed to carry twice the dead and test load. If the floor supporting the safety platform is not capable of carrying the test load and collapsed structure then it, in turn, must be propped from an adequate support structure capable of carrying such a load. The gap between the safety platform and the tested structure must be sufficient to allow the structure to deflect and for the insertion of dial gauges and other movement measurement instrumentation. The contractors must comply with the Health and Safety at Work Act and the area below possible collapse must be closed to access. Instruments should be read, from a safe distance either by telescope or theodolite or by electronic means.

Structural inspection

The structure is likely to have been inspected before testing and material, construction, wear and tear and other defects noted. The presence of defects or the efficiency of remedial measures may be the reason for testing. The behaviour of the defect or its repair should be noted in testing.

Predicted behaviour of structure under test

The engineer should calculate the predicted deflexions and load capacity and estimate the behaviour (and the ultimate load when testing to destruction and mode of collapse). Where, as is often the case, it is difficult to be precise in estimates, the results of the initial load increments should be carefully examined to check the accuracy of the predictions and the need for extra or amended measuring instrumentation. Time-dependent and temperature variations together with their effects on structural movement during the test must also be considered.

Test arrangements

Measuring equipment and instrumentation should be calibrated both before and after the test, be sufficiently sensitive to record small movements and have a capacity to measure twice the predicted maximum deformations. Measurements should be made by remote methods since manual measurements could disturb other instrumentation and can be dangerous to personnel in the event of structural collapse. The test load should be applied at a steady rate without causing impact or vibration.

12.7. Test procedure

Loading

The test load should, as far as possible, simulate the service conditions of the structure. If a beam in use would have to support a uniformly distributed load then the use of a point load at the centre of the span to obtain an equal bending moment would not create the same shear stress conditions. It might, however, be easier to test load a column eccentrically when in actual use its main load would be concentric. Any safety platforms or propping should not provide lateral or other restraints which would not exist in service. Where the loads and restraints cannot be simulated then the effect on the test results should be determined by the engineer.

The test load can be iron weights, sand bags, stacks of bricks, etc., but these have to be weighed and placed in such a manner as to prevent any chance of arching action occurring. A more convenient and economical method is to use water pumped into plastic bins.

Test loads can be applied by jacks reacting against another part of the structure. If the jack cannot be accurately calibrated then load cells must be inserted to measure the pressure. Jacks can only exert point loads; a line of jacks, acting through a stiff steel beam placed between the jacks and the structure can simulate a line load, but to simulate a uniformly distributed loaded area is difficult and can be inaccurate.

Loads must not be dropped or carelessly dumped on the structure as this would cause impact loading. The removal of load must also be done smoothly. The rate of loading should, as far as possible, simulate that likely to occur in use and the number of increments (and decrements on unloading) should be sufficient to allow a number of intermediate readings to be taken so that a better understanding of the structural performance is gained. Load/deflexion graphs are usually linear, and where the graph behaves non-linearly the increments should be decreased and extra caution taken with loading. Should there be a relatively high increase in deflexion for a small increase in load the structure may not recover (and this should be checked by removing the load) and may be heading for collapse. The engineer will have to decide whether to abandon the test or continue to load to destruction.

If cracking, spalling of concrete, crushing of mortar or other signs of distress become serious the engineer should either proceed with extra caution or even abort the test. Load/deflexion readings should be taken both in loading and unloading the tested structure. It is important therefore to ensure not only adequate instrumentation, but also that the instruments can follow all likely structural movement and that they are so positioned that they can be read safely whatever occurs.

Measurement of load
The accuracy of the measurement of the load is obviously important and should be within + 2%. It is relatively easy to do this with water by measuring the head and it is also simple with iron weights of known weight. It is not so straightforward when using sand, bricks or blocks. Pallets of bricks and blocks should be weighed before placing and sand, which can change weight with water content, should be checked for dry weight and variations in moisture content. When using jacks or hydraulic rams, in addition to their calibration it is advisable to confirm the pressure with load cells or proving rings.

The magnitude of the maximum test load depends on

- the design service load

- the strength and condition of the structure and its materials

- the predicted collapse load and the results of a collapse.

The maximum test load should be left in position for at least 24 hours (or that recommended in codes of practice) to determine whether movement is still continuing. If there has been further movement then measurements should be taken daily until, for all practical purposes, the movement has ceased. (Creep of concrete and moisture movement of brickwork, for example, can take years.) Similarly, daily readings of recovery should be taken on removal of load, again until, for all practical purposes, the recovery is complete.

Measurement of structure
Cracks, signs of movement or other forms of distress should be monitored during each increment of loading and unloading. Cracks occur more commonly in concrete, brickwork and blockwork. Where these either have occurred or are likely to in the test, it is helpful to whitewash the crack area to ease identification and measurement. Felt-tipped pens are useful to mark the ends of cracks at each increment of loading and, with a different colour pen, each decrement. The length, position and width of crack should be surveyed. On site crack widths are probably best measured with hand microscopes. Cracks in welds can be identified by using penetrating fluorescent dyes or by painting the crack area with a mixture of oil and talc. If the crack appears on loading, or worsens seriously, the test should be stopped and the cracking checked to determine the likelihood of structural failure. Painting areas of likely high stress concentrations with brittle lacquer can give early qualitative appraisal of possible distress.

Stress can be determined from strain measurements. While this is usually essential in testing prototypes and is sometimes needed in testing to destruction, it is not commonly necessary in acceptance or behaviour testing. Strains can be measured by Demec gauges or electrical resistance strain gauges. Such measurement is best done by trained laboratory personnel.

The most common form of measurement is to measure deflexion with dial gauges or linear variable transducers in the direction of the applied test load. The instruments must be mounted independent from the test element and protected from disturbance. If the instrument supports are likely to move during the test the movement must be measured and permanent datum reference points should be established so that checking can be carried out.

Select bibliography
1. British Standards Institution. *Design loadings for buildings*. BS6399, BSI, London, 1984.

2. Building Research Establishment. *Static load test of building structures.* Information Paper IP/89, BRE, Garston, 1989.
3. Garas F. K., Clarke J. L. and Armer G. S. T. (eds). *Structural assessment — the use of full and large scale testing.* Butterworths, Guildford, 1987.
4. Institution of Structural Engineers. *The appraisal of existing structures.* IStructE, London, 1980.
5. Institution of Structural Engineers. *Load testing of structures.* IStructE, London, 1989.
6. Institution of Structural Engineers. *Load testing of structures and structural components.* IStructE, London, 1989.
7. Jones D. S. and Oliver C. W. The practical aspects of load testing. *The Structural Engineer,* 1978, **56A**, No.12, pp. 347–353; Discussion, 1980, **58A**, No.8, pp. 251–253.
8. Menzies J. B. Load testing of concrete building structures. *The Structural Engineer,* 1978, **56A**, No.12; Discussion, 1980, **58A**, No.8.

13

Unusual materials and structures

13.1. Introduction

Unusual materials and unusual structures are topics which sometimes bear a close relationship to each other, for example unusual materials often require special structural treatment; on the other hand unusual structures are often built of conventional materials. This chapter deals firstly with unusual materials and comments on any special structural application which is appropriate to them. Secondly, it gives some general comment on the appraisal of unusual structures.

13.2. Unusual materials

Historically, building materials have comprised whatever happens to be locally available or could be easily made from substances near at hand. Local stone for masonry, timber if plentiful for frames and cladding, clay for bricks and even ice for igloos and straw for roofs. The use of local materials is still common even in modern building although 'local' now denotes a wider area due to the expansion of transport. Thus timber is the prominent material for domestic buildings in North America, while Britain's wealth of clays enables it to make good use of brickwork. In the more arid parts of the world which have neither timber nor clay in good supply, concrete incorporating local aggregates is used.

A material which appears unusual to the person appraising the building may not necessarily be at all unusual to the locality. Unusualness is subjective and it is often quite easy to gain helpful local knowledge of traditional materials.

For buildings dating from the first half of this century and before, it would be unusual indeed to come across a material which is not treated at some length in other parts of this series of guides. More likely would be the instance of local variations of materials or techniques which would require local investigation.

After the 1950s new materials entered the construction industry, including a completely new family of materials — polymers or plastics. Important developments occurred in higher strength steels, aluminium and later stainless steel became more commonly used. There were considerable improvements in the general quality of concrete and a considerable expansion in the use of prestressed concrete. Many of these developments had their main impact on the non-structural parts of buildings and of their structural influence much is dealt with in other parts of this guide. Nevertheless there remain a number of unusual materials which do not fit into the mainstream and which have been used in a structural or semi-structural mode. They include reinforced plastics, fibre-reinforced cement and ferro-cement.

13.3. Reinforced plastics

As the common name 'plastic' implies, most polymers tend to suffer long-term yield under quite low stress. They also have a low modulus of elasticity. Both these defects can be reduced by reinforcement, usually with fibres of higher modulus and much reduced creep. Various combinations of polymer and fibre reinforcement have been tried but the only combination to emerge in volume in the construction industry has been polyester resin reinforced with glass fibre or GRP (glass-reinforced polyester). For specialist applications of high stress, carbon fibre reinforcement and sometimes epoxy resin is used, but both are more costly than GRP.

GRP is well known in the boat building and vehicle body industries but it is also used, often to striking effect for building cladding, roofing and occasionally for the structure itself. Quite large cladding panels can be made — some up to 10 m × 2.5 m

Solid masonry? Not quite; the door surround and arch are in GRP

have been used — so the structural engineer would be involved in determining their strength and rigidity in respect of wind and roof loading.

In the conventional concept of strength, GRP can be made very strong indeed, over 1000 N/mm^2, but this value will vary enormously with the type and amount of reinforcement incorporated. Immediate or short-term strength is therefore rarely a problem and the more common problems with GRP are long-term creep under sustained stress and its lack of rigidity.

In volumetric terms, polyester resin is expensive so it is nearly always used in thin sections. Skilled designers will have realized the importance of curvature, especially in two dimensions, in achieving rigidity with laminar or shell structures. It is easy and inexpensive to achieve quite complex curved shapes in the manufacture of GRP and the best designs will show plenty of curvature and hence, in all probability, sufficient rigidity for their purpose. Problems arise where the designer has, for some reason or another, preferred flat or almost flat surfaces, and has had to achieve the required rigidity by installing ribs or sandwich construction behind the main surface. Such additions can become delaminated or produce rippling and distortion due to differential thermal expansion of the different surfaces involved. Long-term creep can show up as rippling and distortion of parts of the GRP unit, especially at connections where stresses are concentrated.

Owing to its inherent elasticity and the thin sections normally used, thermal stresses are not usually a problem with the GRP itself provided the designer has made proper allowance for movement. Nevertheless GRP does have a high coefficient of thermal expansion and problems are sometimes encountered from this source at joints and fixings.

Compared with many other construction materials GRP is very durable. It is virtually impermeable, non-corrodable and non-erodible. The polyester resins can, however, break down chemically very slowly under ultraviolet light. The rate of deterioration depends a great deal upon the actual formulation of the resin used. The deterioration occurs on the exposed surface and, when necessary, it can be cured by recoating the surface with a suitable polymer. A number of GRP buildings have performed satisfactorily for 30 years or more and predictions can be given of at least 50 years of useful life before first maintenance, for well designed and manufactured elements.

GRP can be recognized fairly easily by its smooth surface which is warm to the touch, usually with a moulded appearance; indeed only cast iron or precast concrete can match the shapes attainable with GRP. The outside surface of the GRP should be examined carefully for cracks, even quite small ones, especially towards the edges and around fixings. The shape of any cracking should be carefully noted and photographed. The geometry of cracking can give useful clues to its cause. Other defects which must be carefully noted are rippling or distortion of the surface and degradation of the surface itself, i.e. loss of smoothness, a powdery surface or even exposure of the underlying glass fibre. Discoloration could be evidence of local deterioration of the surface. Of particular importance is careful examination of the joints and fixings between GRP elements. More than half the defects usually found in GRP structures are related to the joints and not to the GRP itself. This is particularly true of the earlier GRP buildings when the need for very high standards of joint design were not fully appreciated.

It is particularly important to examine GRP elements for exposed glass fibres which can occur at edges, in deep cracks or in cases of severe surface degradation. Although the glass fibre is not in itself hygroscopic, moisture can migrate along the interface of the glass fibre and the surrounding resin and cause deeper degradation of the material.

When defects are detected and are considered serious they should be referred to a specialist in GRP repair. Minor defects of a cosmetic nature can usually be repaired fairly easily provided care is taken in colour matching.

GRP is combustible and in that respect can be broadly compared with timber. In the grades used in building it is a long way from being highly inflammable and in the forty years or so

These inclined walkways at Heathrow Airport are typical of the use of GRC, of which there is much at Heathrow; but in fact these walkways are in GRP, distinguishable by its glossy surface

113

it has been used in buildings it has not gained a reputation in practice as a particular fire hazard. Nevertheless in many lay minds GRP is categorized as a plastic and therefore as a fire problem. An appraiser might therefore be required to report on the fire risk of a GRP element in a building and in that case the total fire situation of the building has to be taken into account.

There are a number of standardized fire tests available to determine the surface spread of flame and the fire penetration characteristics of GRP. Expert help should be sought if there is thought to be a particular fire problem.

13.4. Fibre-reinforced cements

Asbestos fibre has been widely used as a reinforcement for cement and the resulting product is very well known. Brittleness of the product has limited its use to semi-structural applications and more recently some asbestos fibres have been revealed as a health hazard. Various other fibres have been tried experimentally and one, glass fibre, has emerged as a usable product. GRC is a patented product which in many ways can be compared with asbestos cement, although the former has much greater ductility and higher strength.

GRC is used for building construction in the form of cladding panels. The cement mortar and fibre mix is normally sprayed into a mould. The external appearance of this product is similar to precast concrete and can be confused with it. The much thinner sections achieved by GRC and its somewhat denser surface clearly distinguishes it. GRC should be as durable as the best quality precast concrete but it is basically a concrete and is therefore porous to some extent. The surfaces are commonly painted but some weather staining might occur.

13.5. Ferro-cement

Ferro-cement is very thin reinforced concrete, with thicknesses commonly of 2 cm or less. It is created by forming first an armature of fine wire mesh and then rendering each side of the armature in turn with cement mortar. Various refinements are found, e.g. with specially woven meshes and mechanically applied or vibrated mortar. The technique in fact predates normal reinforced concrete, having been invented in 1840. The material has been widely used in boat building because of the ease in which it is formed into complex curved shapes and, in the developing countries, for structures such as water tanks and small buildings. It was first used on major structures by Luigi Nervi in the 1940s and 1950s and much of his ferro-cement work is quite famous, although most is in foreign countries. Research and development on this material is going on in the Far East and in Russia where ferro-cement roof units are mass produced. It is therefore overseas that the engineer is most likely to meet the material. Despite the thin sections, ferro-cement is used quite successfully for water-retaining structures; any cracking that does occur is very fine because it is controlled by the fine mesh reinforcement. Faults that can occur are mainly due to inadequate working of the mortar into the mesh, so hollowness and the cracking that is symptomatic of hollowness is the main thing the engineer should look out for.

13.6. Unusual structures

It is difficult to define what constitutes an unusual structure as in one sense almost every building structure is unique in some way and therefore 'unusual' yet almost every structure on earth (and most of those in space!) are built on structural principles with which every engineer will be very familiar. This latter point provides a clue to how the engineer should approach a structure which appears unusual. First of all he should follow the advice given in Chapter 2 of this guide, particularly regarding how a structure works. It is most important to satisfy oneself that there is a clear and consistent path through the structure for all loads — including lateral loads and those self-induced by expansion or settlement — to be accommodated or transferred to the foundations.

Some unusual structures involve structural forms such as arches, domes, cones, vaults and folded plates which require lateral restraints for their stability. The basic mathematics for such structures is well established but, unfortunately, the everyday work of a modern structural engineer concerns beams, slabs and columns almost exclusively. It is well known that arches and pitched rafters require horizontal restraint but many structural defects occur through the inadequacy of those restraints. Signs of spreading arches and rafters should be looked for. Domes and cones are uncommon but again circumferential restraint must be adequate.

Barrel-vault roofing in reinforced concrete enjoyed a period in vogue in the 1930s to 1950s. The stress distribution in such structures is quite complex but most were designed by competent engineers and rarely show structural problems. Variations of vaulting occur in the hyperbolic paraboloid form in concrete and timber but their very existence betokens the work of a specialist designer. Even so, a careful inspection should be made for signs of overstress, particularly at points of support, and for signs of lateral movement or settlement. In some concrete structures of this form differential settlement can induce high stresses.

In more recent times, unusual structures have become more self-evident with the exposure, often with striking architectural effect, of the main elements, be they compressive, tensile, torsional or flexural. Where cables or rods are used as exposed tensile members, careful examination should be made of the protective coatings and the means of excluding moisture. Such structures often involve quite high stresses at connections and bearings. Any cracking surrounding such points should be carefully considered.

A new generation of unusual structures has emerged in the last decade or so in the form of fabric structures. The development of high strength polymer fabrics enables them to be used suspended from more usual structures — a development of the tent — or supported by slightly increased internal air pressure. Sometimes a combination of support systems is used. The fabric can only be used in tension, but considerable tensile loads can be borne so that the fabric itself forms an intrinsic part of the total structure. Such techniques have not been in use long enough for a working vocabulary of defects to emerge but if one is to be appraised, then a diligent pursuit of the structural principals on which the building is designed should lead to a sufficient understanding of the defects which the engineer should be alert to. If that is not the case, expert advice should be sought. In the case of air-supported structures, attention should be given to the state of affairs that might be brought about by a failure of the inflating machinery or by a fire. Many air-supported buildings have secondary means of support in the event of air pressure failure and in many other cases the collapse of the envelope would be quite slow. Nevertheless the safety of the occupants of that building needs special consideration. There are of course hundreds of unusual structures — from tree houses and mobile homes to offshore oil platforms and underground dwellings, from space laboratories to submarine chambers — all of which can in a general sense be called buildings. Clearly the more bizarre the example the more likely expert help is to be available but one should take comfort from the fact that there is no conceivable structure which does not follow, in some way or other, the elementary principals of applied mathematics.

Select bibliography

1. Building Research Establishment. *Durability and applications of plastics.* BRE Digest 69, BRE, Garston, 1977.
2. Building Research Establishment. *Cellular plastics for buildings.* BRE Digest 224, BRE, Garston, 1979.
3. Building Research Establishment. *Glass reinforced cement.* BRE Digest 331, BRE, Garston, 1988.

14

Historic buildings

14.1. Introduction

There is an increasing awareness that historic buildings and their settings are an irreplaceable part of our environment and that those entrusted with their appraisal and repair should employ sympathy and care in what they do and operate within a legal framework, the aim of which is to preserve these buildings for prosperity. Notwithstanding any statutory protection a building may or may not have, the appraising engineer owes it to himself and his profession to be sensitive when dealing with alterations to all buildings, and especially to historic ones.

14.2. Statutory protection

An historic building may be protected either by scheduling or by listing. Ancient monuments are scheduled under the Ancient Monuments and Archaeological Areas Act 1979 by their inclusion in lists prepared and published by the Secretaries of State for the Environment (in England), for Wales and for Scotland. The object of scheduling is to ensure that no works affecting an ancient monument are carried out without first obtaining statutory consent. Listed buildings are protected by the planning control activity of the local authorities, although the lists themselves are prepared and published by the Secretaries of State mentioned above. The effect of listing is to make it necessary for consent to be obtained before carrying out works of demolition, alteration or extension in a manner which would affect the character of a building of special architectural or historical interest.

In England and Wales listed buildings are classified into three grades to indicate their relative importance:

- grade I — buildings of exceptional interest

- grade II* — particularly important buildings of exceptional interest which warrant every effort being made to preserve them

- grades II and III — these are large categories of 'interesting' buildings which, although listed, are not subject to quite so rigid a control as the higher grades.

Most buildings built before 1840 which survive in anything like their original condition are likely to be listed. Between 1840 and 1914 only buildings of definite quality and character are listed, and since then selected buildings thought to be of high quality or exceptional interest find their way on to the list. A listed building includes any man-made object or structure fixed to it or which was within its curtilage before June 1948 such as internal and external features, boundary walls, out-buildings, etc.

In Scotland, such buildings are graded on a different basis.

- category A — buildings of architectural or historic interest or fine, little-altered examples of some period or style which are of national or more than local importance

- category B — buildings of primarily local importance or examples of some period or style which may have been altered in a minor way

- category C(S) — good buildings which may be considerably altered and other buildings which are fair examples of their period, or in some cases buildings of no great individual merit which group well with others in category A or B.

Buildings which are neither listed nor scheduled may be subject to temporary listing if a local planning authority considers that a building of listable quality is in danger from demolition or alteration. The authority will in such cases, serve a building preservation notice on the owner which has the effect of listing the building for six months while formal listing is contemplated.

Some local planning authorities keep 'local lists' of buildings of local architectural or historical interest. Even though such lists have no legal significance, it is nevertheless advisable to work closely with the authority.

A group of buildings may be protected as part of a conservation area which has been designated by the local planning authority or by the Secretary of State. Individual buildings in the area may not be of listable quality themselves but the permission of the local planning authority is required for any significant alterations or demolitions, even extending to the trimming, cutting down or uprooting of any tree over 76 mm in diameter.

Areas of archaeological importance may be designated in England and Wales by the Secretaries of State or, subject to their confirmation, by a local authority. Anyone intending to disturb the ground (including flooding or tipping) in such an area must serve a six week operation notice on the local authority which has the power to delay the work by up to six months to permit archaeological investigation.

14.3. Consultation

Although initial consultation should be through the local authority, the Historic Building and Monuments Commission should be notified, and its comments and advice invited, if it is proposed to carry out any of the following works to a grade I or II* listed building:

- demolition

- alteration, externally or internally, or extension in any way which would affect the character of the building

- repairs or other works involving the loss or replacement of historic fabric

- any proposals which might affect the setting

Such consultations should take place as early in planning as possible. The following list gives some idea of the matters of detail which concern the Commission. The list is not comprehensive and in cases of doubt the Commission's advice should be sought.

- Brick and stone renewal and repointing
- Re-cutting or re-working of decoration in stone, brick, plaster, timber, lead, etc.
- Timber repairs to roofs, wall framing and floors
- External cleaning and external redecoration where a change is proposed
- The replacement of doors, window frames and lead water pipes
- The replacement of internal fittings such as panelling, door furniture, etc.
- The re-decoration of architecturally or historically significant rooms where a change is proposed
- The installation or renewal of main services or air conditioning equipment.

Proposals to demolish or to alter or extend grade II listed buildings are normally dealt with by local planning authorities, although in the case of demolition the Commission should be consulted. However, in Greater London proposals which would affect the settings of grade II listed buildings should also be notified to the Historic Buildings and Monuments Commission.

Any work affecting scheduled ancient monuments and known archaeological remains — including demolition, removal or repair of any part, alteration or addition, and any flooding or tipping operation on the land — must be referred for approval to the Historic Buildings and Monuments Commission. In the first instance approval in principle should be sought from the Commission and later followed up for detailed approval when full proposals have been produced.

14.4. Principles affecting appraisal and repair

A proper appraisal can only be made if the building's history and materials are understood. It is likely that some papers or sketches exist whereby the history of the building may be traced, and may reveal earlier significant construction. Knowledge of previous repairs or alterations will suggest likely defects to look for and will help in their interpretation. An understanding of historical construction will also assist when assessing visible defects and prediction of hidden ones. As a general principle, maintenance repairs should be carried out using the same materials and techniques as for the original work. When this is not practicable, modern remedial methods (e.g. the use of epoxy resins) may prove to present clean and cost-effective solutions by reducing the amount of cutting out required. Where possible however, repairs should be reversible so that those with responsibility for maintenance and repair in the future can take joints etc. apart if necessary. It is well worthwhile taking advice from specialist craftsmen (e.g. stonemasons) before applying modern techniques to old structures. Many apparently inert materials have peculiar characteristics of their own.

The cost of access may be the predominant one in any scheme and funds may be limited. In such cases it is usually more cost-effective to concentrate on those features most at risk than to spread the available finances thinly over the whole of the building structure.

The Parthenon, Athens, was recently the subject of what must have been a fascinating structural appraisal.

The cumulative effect over the years of a series of minor changes can be very damaging to the historical character of a building. As much planning as possible should therefore be done before work starts on site as otherwise what start off as repairs could develop into what might be regarded as alterations in the course of the actual work.

14.5. Principles applying to alterations

Substantial changes may be acceptable in a much altered building of minor historical interest, but not in an unchanged building of great original value. However, past changes can sometimes contribute to the interest and value of buildings. It should not be assumed that interiors are of less intrinsic value than the exterior; the contrary may well be true. Minimum change to the original structure should be high on the list of priorities and, on buildings of importance, alteration should, so far as is possible, not entail irreversible changes. Further guidance on more detailed aspects relating to the alteration of listed buildings is contained in a technical digest (Appendix IV to Department of the Environment circular 8/87) prepared by the Historic Buildings and Monuments Commission (5 in bibliography).

14.6. Repair problems

Few materials or elements of construction used in historic buildings were impervious to moisture so a high rate of ventilation through loose fitting doors and windows and flues was necessitated. Changes to such construction carries the risk of disrupting the equilibrium achieved by the original builders. In particular, surface treatments designed to weatherproof building elements must be regarded with caution. They often only tackle the visible effects of deterioration and can themselves cause quite different problems.

Much more timber was used in buildings in the past than in recent times and even apparently solid masonry walls can contain built-in timber in the form of lintels, joists, spreaders and bonding timbers. The rusting of embedded ironwork can disrupt the building fabric as it expands. Corrosion of straps and ties can be dangerous as they may burst suddenly, and it is therefore important to search out where these are likely to have been used.

Traditionally, brickwork and masonry were built with relatively soft lime mortar which enabled the walls to move without cracking the bricks or stones. A further benefit was that water could dry out of the walls through the joints, thus affording some protection to the masonry units from efflorescence and frost damage. Repointing with harder materials should therefore be avoided.

The traditional approach to structural movement was also completely different. Much historical internal detailing, such as cornices and architraves, accommodates movement and disguises irregularities. Repairs should not attempt to restrain such movement unless the cause is such that it will eventually lead to unacceptable damage.

The temptation to repair all cracks in old structures should be resisted. Many cracks are beneficial as they have in the past relieved stresses, so preventing more serious damage, or they allow the structure to move safely to accommodate seasonal and diurnal changes.

Many walls or pillars of old buildings may give trouble because of the use of poor quality cheap materials behind an attractive face of durable material. This filling, in turn, may be inadequately bonded to the surface material and separation failures can be initiated, aggravated by water penetration and movement due to other causes.

There was little understanding of foundation design in the past, resulting in major differential settlements of different parts of structures or the setting up of large shear stresses. Care should be taken to account for the releasing of such stresses.

14.7. Execution of work

It is most important that only firms which can show that they have carried out similar work satisfactorily should be invited to tender. Similarly, supervision and site control must be in the hands of experienced personnel. On historic buildings, the responsibility cannot be left to the contractor for determining the extent of taking down, cutting back or opening up. Rapid decisions, however, must be available for the contractor regarding how far to go at any point and how to cope with hidden defects. A thorough record of the work must be kept, not just for contractor control but to document it for the benefit of later investigators. Photographs taken before, during and after the work is carried out are an invaluable record.

Attention must be paid to protection of the original fabric and historic contents and emphasis must always be on prevention of damage rather than acceptance and making good afterwards. Important temporary works, such as strutting and other temporary structural support, should be to the design of the engineer and not delegated to the contractor. Scaffolding can easily damage the fabric and rust staining from scaffolding which may stand for a long period of time should be avoided. Care must be taken, however, not to absolve the contractor of his responsibilities. Careful specification and a rigid system of approving the design and installation of temporary works is necessary.

The position of inserted fixings should be agreed and recorded on the drawings in advance and only allowed if inconspicuous. Special care needs to be given to the durability of even the smallest items introduced into the fabric such as screws, bolts and reinforcing steel. The standards of protection of mild steel which are generally sufficient in normal cases may not be adequate for the life expectancy required in an historic building. The engineer should, of course, be aware of the consequences of using dissimilar metals in contact with each other.

14.8. Financial considerations

The problems of project cost control are obviously greater where the full extent of work cannot be determined in advance. Limited funds may be available and if temporary, or even permanent, suspension of the work is a possibility, the phasing should be such that it allows wind and weatherproofing to be carried out quickly. It follows that the contract should be phased so that only minimal areas are open to the weather at any time.

There are a number of sources for financial assistance for the conservation of historic structures. Local authorities, the Architectural Heritage Fund and some private trusts may be approached and, for the more important projects, the National Heritage Memorial Fund. The chief source of financial assistance in England is English Heritage who operate three main grant schemes covering buildings considered to be of outstanding national interest (generally grade I or II*), ancient monuments, work to preserve and enhance the character and appearance of a conservation area (in close co-operation with local authorities) and, more rarely, for purchase (usually by a local authority) of property for the express purpose of its preservation. The detailed rules for eligibility under these schemes vary slightly to reflect their objectives. Conservation area grants, for instance, are limited mainly to work on the external envelope of the building, whereas other grants are concerned with total repair. Grants are rarely given retrospectively and are normally available only for schemes of major repairs or restoration; not for minor works and never for alterations, improvements or other new work, however desirable this may be to keep the structures in use. Grants, moreover, cover only a proportion of the repair costs and applicants will be expected to demonstrate that they cannot reasonably meet the cost of work themselves. Those administering a grant scheme are likely to require assurance both as to the financial viability of a project and that the work will be executed to an acceptable standard.

Select bibliography

1. Ancient Monuments and Archaeological Areas Act 1979. HMSO, London.
2. Brunskill R. W. *Illustrated handbook of vernacular architecture*. Faber & Faber, London, 1987.
3. Cambridgeshire County Council. *Cambridgeshire guide to historic buildings law*. Cambridgeshire CC, 1988.
4. Helme D. (ed.) *The conservation handbook*. Property Services Agency, Croydon, 1988.
5. Historic Buildings and Monuments Commission. Appendix IV to DOE Circular 8/87, Technical Digest, Department of the Environment, London, 1987.
6. Historic Buildings and Monuments Division, Scottish Development Department. Memorandum of guidance on listed buildings in conservation areas. Scottish Development Department, Edinburgh, 1987.
7. Hume I. J. Assessment, monitoring and temporary support of historically important structures. *Proceedings of Institution of Civil Engineers Conference on Conservation of Engineering Structures*. Thomas Telford, London, 1989.

8. Lemaire R. M. and Van Balen K. (eds). *Stable — unstable — structural consolidation of ancient buildings.* Leuvan University Press, 1988.
9. Mitchell E. *Emergency repairs for historic buildings.* English Heritage,
10. Parnell A. C. *Building legislation and historic buildings.* Architectural Press, London, 1987.
11. Powis A. R. *Repair of ancient buildings.* Society for the Protection of Ancient Buildings, London, 1981.
12. Scottish Development Department. *New provisions and revised guidance related to listed buildings and conservation areas.* Circular 17/1987, Scottish Development Department, Edinburgh, 1988.
13. Sharman E. J. Legislative framework and financial assistance. *Proceedings of Institution of Civil Engineers Conference on Conservation of Engineering Structures.* Thomas Telford, London, 1989.
14. Society for the Protection of Ancient Buildings. Various useful technical pamphlets and information leaflets.
15. Suddards R. W. *Listed buildings — law and practice.* Sweet and Maxwell, London, 1982.

15

The report

15.1. Introduction

Structural appraisal reports may have to be used and understood by lay readers, may be used as legal evidence, and can have serious cost and planning implications for the client. They also tend to cover a wider range of topics than most technical reports and deal with defects in construction or design, deterioration, collapse or accidental damage, change of use of the structure and its future safety and durability, inspection for the prospective buyer or change of insurance rating, and increasingly, to settle legal disputes.

Although engineers tend to be more visually orientated and numerate than literate, this is not necessarily a disadvantage because sketches, photographs, graphs and similar visual aids are more readily understood than long verbal descriptions. It is also necessary to appreciate that there is a considerable difference between spoken and written English.

15.2. Basic principles

Drafting the report

It is rare for an author to write a final and polished report at the first attempt. Reports need drafting, re-drafting, reviewing and assessing by colleagues, re-ordering, revising, checking and amending, just as does engineering design. It is not uncommon in practice for important reports to go through six drafts with the penultimate drafts having the meaning of every sentence, sketch, graph, etc. critically examined. It follows that word processors are essential.

Simplicity and clarity
Since the report will be read by non-technical clients (councillors, civil servants, lawyers, businessmen, etc.) it must be clear, simple, brief and easily understood. Where technical terms are unavoidable, and this is common, consideration should be given to explaining them. It can be useful, as a trial run, to have the report read by a layman in the office (who can guarantee confidentiality) before issuing it to a client.

Technical accuracy, precision, and interpretation
No engineer would be happy to issue or use a working drawing that was inaccurate, imprecise and open to misinterpretation; the same criteria of accuracy, precision and correct interpretation apply equally to reports. It should be self-evident that technical accuracy is imperative. When, for any reason, total accuracy is not possible it is essential to state why there could not be such accuracy and how comments on the available facts have to be qualified. Precision should not be confused with being concise — precise means definite and exact, not approximate and not subject to reservation. Interpretation is the explanation of the facts and observations, results and the like, so that the reader clearly understands.

Readability
The report must be logical, objective and reasonable and not emotional, biased or subjective. It should have sensible continuity so as to be easy to follow (this sounds obvious but too often the report writer is too busy with the trees to see the wood). It must be relevant to its purpose and must not omit anything of importance. It is essential that it is based on factual data and that subjective comment is kept separate from the facts as far as possible. Where it is unavoidable that there are reservations or limitations about the survey details, methods, tests or techniques or the interpretation of survey results, then these must be honestly stated and explained.

Recommendations and conclusions

It must be made clear that the recommendations are based on the facts and experience, and where there are alternative recommendations (as there often are) the advantages and disadvantages must be carefully defined. Further, the engineer should state his preferred recommendation and not leave the client with a bewildering array of options. The client may not accept the engineer's preferred recommendation, for often there are social, political, financial or other factors which could be of more importance than the engineering alone.

The conclusions must be definite (not implicit), unambiguous, comprehensible and positive. The conclusions must only be qualified when adequate information was unavailable or the survey and tests were inconclusive.

Checking final draft

A report should not be issued without a check on the presentation as well as the technical content. The pointers given above should be checked for their application in the report. An experienced engineer, not involved in writing the report, should be asked to examine it critically, not for its technical content but for its style, by asking the following questions.

- Is it simple and clear, accurate and precise?

- Does it read well without confusion?

- Are the recommendations clear and are the conclusions helpful, coherent and comprehensible?

15.3. *Report writing procedure*

The procedure set out here applies to all structural appraisals. The size of the project is no criterion of its importance nor its need for care, procedure and skill and experience. The report for the historic Albert Dock ran to 7000 pages of text, 3000 sheets of drawings, 800 photographs and 2000 sheets of test reports. A report on the bay window of a Victorian terraced house consisted of just six pages. The basic investigation, procedure and care and report writing was similar in both surveys. Some surveys can be dauntingly vast and complex so procedure is vital and the following procedure has been found to be efficient for both large and small surveys.

Collect data, categorize it and write first draft
A report can contain a great amount of information and it is essential to consider carefully just what data is necessary before starting. The data collection will often require amendment as the work proceeds and, in some cases, early investigation can show that due to the unacceptably high cost of the required remedial measures, the client should be advised to consider terminating the project. On major appraisals it is good practice to report regularly to the client on progress and cost implications.

The mass of collected data needs organizing into appropriate categories and sections, and this should be started in the office as the data arrives. Obvious categories are soil conditions, foundation faults, structural frame deficiencies, flooring and roofing problems, cladding deterioration, etc. The sections on foundations, for example, would break down into types, condition, adequacy, need for any remedial measures, test results and similar factors.

Such categorization not only speeds and improves the final report but it is extremely helpful in guiding any necessary further detailed investigation.

Edit first draft and decide on methods of visual presentation
If the engineer writing the preliminary draft is not the investigating engineer there should be discussion between them on matters such as accuracy, completeness and interpretation. When the first draft is typed, preferably on a word processor, there are usually sections that very obviously need improving. As more data is collected the need for sub-sections, headings and cross-references becomes apparent. Typical cases are where all steelwork has been attacked by a highly corrosive environment or concrete is suffering from the effects of calcium chloride additives. While sections on the material are necessary, so too are separate sections dealing with the attack and its effects. When symbols and abbreviations (Z for section modulus, γ_m for partial safety factor of materials, PRC for prefabricated reinforced concrete, etc.) are used, particularly for the first time in the report, they must be given their meaning. If it is necessary to use many symbols and abbreviations then a list should be included after the contents list.

Obvious visual aids are the small- and large-scale Ordnance Survey maps to locate the site; plans, sections, elevations and photographs of the structure; large-scale details, isometric sketches and photographs of elements and faults; photographs of site and laboratory tests, tabulation and graphs of test results; and histograms, charts and similar visual representations of facts. All must be captioned, numbered, indexed and mentioned separately in the contents list.

Penultimate draft
The penultimate draft must be checked for

- completeness — no important information omitted

- clarity — lack of duplication; inclusion of irrelevant information; mathematical, spelling and grammatical accuracy; correct numbering and sequence of sections, figures, tables, graphs, photographs, etc.

- continuity and logical progression and interpretation

- presentation — main sections should start on a fresh page; margins should be constant; points should be enumerated; pages numbered; figures, photographs etc. should be inserted adjacent to reference in text; the use of capitals, embossing, underlining sub-headings and similar aids to reference.

When this has been completed it should be passed for comment to colleagues who should adopt a critical and 'devil's advocate' attitude. Their comments should be marked on the typescript, preferably in red ink so they are easy to see.

Final draft
This should be checked yet again before issue. Photocopying should be checked for fading and centring. Before binding copies, checks should be made for omission of pages.

15.4. Factors affecting quality of report

Restraint of time and funding are also major factors. Time must be allowed for writing just as it is for design and detailing. While there is seemingly never enough time for design there is usually even less for appraisal reports. Clients are normally unaware of the time needed for investigations and can easily become impatient waiting for the report, so programming is essential. Similarly, few clients have the expertise and experience of costing investigations and if funding is restricted and this affects the quality of the appraisal then this must be made clear in the report.

Sometimes specialists have to be consulted (e.g. regarding uncommon materials such as cast and wrought iron, modern plastics and composites, ancient timber piling, glue analysis of glulam timber, etc.) and this too should be reported with any delay the consultation necessitated in producing the report.

When there are site difficulties (e.g. lack of access, buried and undetectable items, restricted access) these should not be allowed to constrain the appraisal to a previously approved time allowance and cost, but should be reported to the client. If 'extra-over' on cost and time is not granted and this affects the quality of the report, then this must be clearly stated in the report.

15.5. Sequence of report

Structural appraisal reports follow the sequence of items common in engineering reports and papers: title page, contents list, brief synopsis, introduction, the main body of the report in sections and sub-sections, conclusions and, finally, recommendations. Description of site, laboratory tests and similar matters which could affect the flow of the report should be given in appendices, but tabulated test results should be included in the body of the report.

Clients will tend to read only the synopsis and recommendations. If repair contractors are involved they will concentrate on the body of the report and the engineers who may follow up the report with remedial measures will pay most attention to the conclusions and recommendations. If the report forms part of an expert opinion in a legal case then the whole of the report will be subject to close scrutiny.

Title pages

This is normally one page giving the engineer's project reference number, the title 'Structural appraisal of. . . [the building, type and use]', the site location or address, the name and address of the client, the name and address of the engineer, the date of issue and whether the report is final, preliminary, interim, confidential, restricted or otherwise.

Contents list
This gives the section, sub-section and appendices numbers, and the title and page numbers. Figure and photograph numbers, their captions and page numbers should follow and finally, if necessary, a list of symbols and abbreviations giving their description should be included in alphabetical order.

Synopsis
This can usually be written only after the report has been completed and checked and constitutes a précis of it. It should preferably not exceed one page in length and be so written as to increase the reader's motivation to read the rest of the report.

Introduction
This section is better left until the body of the report has been written, when it is easier to give an overview of the report as a whole. It should start with a description and definition of the client's brief and the reasons for it. If the brief had to be amended during the investigation then the causes and details of the changes should be described. This should be followed by a brief description of how the work was carried out and the time scale. Finally, the scope of the work should be outlined. Any limitations imposed on cost and time which affected the quality of the investigation so as to cause any qualification of the conclusions and recommendations should be made clear.

Main body of the report
This is where breaking down of the whole into sub-sections is most necessary, for this can consist of a large number of separate problems, tests, calculations, etc. It should clearly and fully define the problem, state how it was investigated, what was investigated and the results of the investigation. Common sub-sections are

- the building — a brief description of the building giving its location, age, use (past and present), general condition, details of important alterations, existing drawings and calculations and any previous tests and investigations

- the investigation — details of all investigations and inspections should be given. Any load tests, tests on materials and similar should be described (aided by photographs and drawings) and the reasons for carrying them out, and any limitations on them

135

should be given. If the methods, descriptions or calculations are lengthy they can be placed in appendices and only summaries included in the body of the report

- assessment — the results of the tests, calculations, inspections, etc., should be given and where possible these should be presented in a graphical or other easily visualized form.

Conclusions

After all the information gathered has been assessed and fully considered it is possible to begin definite, well researched judgement and conclusions. It is essential to keep facts separate from comment and opinion. The conclusions must be based on the findings discussed in the report. If it is necessary to discuss the tolerance of accuracy of tests and methods and their effect on the findings then it is advisable to state such limitations.

Conclusions should be related to the requirements of the brief but are likely to state clearly whether or not the building is adequate for its present use or a proposed change of use; is in need of maintenance or repair; the building's possible life span; if it is inadequate and beyond repair; and if it is unsafe and should be evacuated. The engineer should not value the property, claim it will last forever, give legal opinion other than as an expert witness, or give any other conclusions outside his professional expertise.

Recommendations

These will 'approve' the building; propose maintenance; advise periodic inspection; suggest demolition, etc. The recommendations must be written so as to be easily understood by lay readers and be based on the conclusions and not be at variance with the rest of the report.

Closing sections

These include appendices, references and an index. Appendices are frequently the test descriptions, methods and reports by specialists in physics, chemistry etc. Where reference is made in the report to codes of practice, the Building Regulations, British Standards, technical papers and the like, the full title, date of issue publication and the issuing organization or publisher, etc. should all be given in the list of references.

Indexes are especially helpful in lengthy and complex reports which may need second readings.

15.6. Presentation

Since reports are likely to be subject to frequent use then merely stapling together a mass of A4 sheets is inadvisable. They should ideally be bound (e.g. in ring binders) and provided with stiff covers to improve presentation and protect the document from wear and tear.

Select bibliography

1. Seeley I. H. *Building survey reports and dilapidations.* Macmillan, London, 1986.
2. Institution of Civil Engineers. *The presentation of engineering evidence.* ICE, London, 1946.

Biographies of the authors

Bill Curtin left school at 13 to work on a building site. By the age of 17 he was a site engineer, building ordnance factories in preparation for the Second World War.

After the war he worked for a newly established small consulting practice headed by Freddy Snow. In 1948 he joined the London County Council, but quickly moved on to lecturing at the Brixton School of Building. In 1951 he was appointed senior lecturer at the Liverpool College of Building where he carried out research into stressed skin timber plates, folded plates, shells, geodesic domes and glulam.

By 1957 he was receiving his first appointments for foundation design and in 1960 he founded his own practice, Curtin's Consulting Engineers. The practice has grown to become a large and highly respected civil and structural engineering consultancy.

Bill Curtin semi-retired in 1976 and returned to research. He took a two year part-time masters degree on plain masonry diaphragm walls, followed by a PhD on prestressed brickwork which he completed in only 10 months!

He served on many committees and working parties, including the Structural Engineering Group Board of the Institution of Civil Engineers, the Construction Industry Research and Information Association's Council and the BSI code of practice committee for structural masonry. His contributions were recognized by several awards for research and innovation and he was twice awarded the prestigious Henry Adams Bronze Medal for original research.

Bob Holland's initial education in the art and science of engineering proceeded concurrently with his training and hands-on experience in most of the trades involved in the structural steelwork and heavy mechanical engineering industries. Service with a Sapper port squadron was followed by experience with contractors, a consultant and a County Architect's department, and six years in a team designing and supervising the construction of a large new sewage treatment works.

He joined the Property Services Agency (PSA) in 1970 and was at one time Head of Structures, where he increased the range of specialist support available to civil engineers working in regional units of the PSA whose duties included the appraisal, repair and maintenance of buildings. During this time his team evaluated competing reinforced concrete repair systems, produced design guides for tall chimneys and masonry walls, and supervised the first successful large-scale use of cathodic protection on a reinforced concrete building structure. He also served on several BSI technical committees and on the ICE Structural Engineering Board.

Two years leading the team responsible for the construction of new prisons led to his appointment as Director of Works for new construction work and large-scale refurbishment at special hospitals and research and training establishments. His final PSA post was heading an 80-strong team of architects, engineers and technicians as a Director of Design. Since leaving PSA in 1991 Bob has undertaken private consultancy work and appeared as an expert witness.

Alec Leggatt spent the early years of his career after graduating in design and site supervision of heavy civil engineering such as hydro-electric and large port schemes. Then a short spell with a specialist contractor in geotechnical site investigation was followed by the realization that creative engineering design was his first real interest. His career as a consulting engineer then began, first with architects in a combined practice for a few years and then for twenty-five years with Nachshen, Crofts and Leggatt until retirement as Chairman in 1989.

Although most of his designs were perforce in conventional steel or concrete he was always keen to exploit innovation in the use of these and newer materials. When GRP arrived he saw it as an excellent material, especially for roofing and cladding. Some of his projects, for example, the Flower Market for New Covent Garden and Sharjah International Airport, were pioneering landmarks in the large-scale use of GRP. A large structure in ferrocement was added to the list. He is the author of *GRP and Buildings* (Butterworth) which was later translated and published in Russia. Now retired, he is actively engaged in arbitration and in voluntary work for the engineering profession in relation to the European Community.

Bruce Montgomery-Smith has specialized in commercial building structures and maritime engineering and made a particular study of contract law, arbitration and contract administration. He has acted as arbitrator and expert witness in a number of cases of building failure and in contractual disputes. He has served on the Council of the Institution of Structural Engineers and on the Structural Engineering Board of the Institution of Civil Engineers.

After graduating he joined Oscar Faber & Partners in London and later in West Africa was in charge of their Nigerian office. Subsequently he moved to Glasgow and was responsible for their work throughout Scotland. He set up his own practice in 1972 to provide a comprehensive professional service in structural, civil, maritime and public health engineering throughout Scotland. The firm has developed a particular specialization in the restoration and adaptation of existing buildings, many of them listed.

He is the author of several published works and has presented a number of technical papers at the institutions and at conferences.

John Moore graduated from Cambridge and London Universities and then remained at Imperial College to attain a doctorate in rock mechanics. After working for the National Coal Board in field trials of prototype equipment and in research in the Rock Mechanics Branch, he joined the Geotechnics Division of the Building Research Establishment (BRE) in 1969.

He worked there on the stability of slopes and deep excavations, particularly developing instrumentation and monitoring the behaviour at full scale of stiff clays and chalk, associated foundations and adjacent buildings, including the underground car park at the Houses of Parliament.

After two years in the Research Management Division of the Department of the Environment involved with research on housing, aggregate and mineral resources, engineering geology and mapping and the construction industry research associations he returned to BRE in 1976 to work on masonry, building integrity and concrete before heading the Structural Integrity Division in 1982. He was responsible then for research on masonry, concrete, structural stability, radioactive waste disposal and multi-disciplinary failure investigations. Particular issues included cavity wall ties, GRC cladding, prefabricated reinforced concrete housing, large panel systems and assessment of structural Eurocodes.

After a short spell as acting Director of the Geotechnics and Structures Group he moved back to the DOE at the end of 1990 to head the Structural Engineering Branch of the Building Regulations Division. Revised requirements for Part A: Structure and a revised Approved document were published at the end of 1991.

Robert W. Turner after being articled with a reinforced concrete company in 1943, served for three years with the Royal Engineers. Employment with consultancies (Ove N. Arup, R. J. James and Andrews Kent & Stone) was followed by seven years in railway civil engineering. After his appointment as Development Engineer with BCSA from 1959 to 1962, twenty-five years was spent in a medium-sized consultancy where he became partner with overall responsibility for engineering matters on many major projects at home and abroad. With the dissolution of that partnership in 1987 the new practice Robert W. Turner Associates was formed, dealing exclusively with forensic work (investigation/litigation work having developed since about 1965). Acting also as consultant to the Mott MacDonald Group since 1988, he continued in forensic work as well as managing various engineering projects.

His forensic work to date includes court appearances as expert witness, surveys and reports on building distress and failure, fire damage etc. at home and abroad.

Institution activities have involved submission of various papers, representation on the Institution of Civil Engineers Structural Board in 1984 to 1987, and being an examiner in 1985 to 1987.

Robert Turner was also a contributing author to *The steel designer's manual.* (Crosby Lockwood) and the *IStructE steelwork design guide*, 1989.

Index